RESPONSIBLE SCIENCE

SCIENCE

Ensuring the Integrity of the Research Process

VOLUME I

Panel on Scientific Responsibility and the Conduct of Research

Committee on Science, Engineering, and Public Policy

National Academy of Sciences
National Academy of Engineering
Institute of Medicine

NATIONAL ACADEMY PRESS
Washington, D.C. 1992

National Academy Press • 2101 Constitution Ave., NW • Washington, DC 20418

NOTICE: The project that is the subject of this report was approved by the Governing Board of the National Research Council, whose members are drawn from the councils of the National Academy of Sciences, the National Academy of Engineering, and the Institute of Medicine. The members of the panel responsible for this report were chosen for their special competences and with regard for appropriate balance. This report is the result of work done by an independent panel appointed by the Committee on Science, Engineering, and Public Policy, which has authorized its release to the public.

This report has been reviewed by a group other than the authors according to procedures approved by a Report Review Committee and by the Committee on Science, Engineering, and Public Policy. Both consist of members of the National Academy of Sciences, National Academy of Engineering, and Institute of Medicine.

Library of Congress Cataloging-in-Publication Data

Committee on Science, Engineering, and Public Policy (U.S.). Panel on
  Scientific Responsibility and the Conduct of Research.
    Responsible science : ensuring the integrity of the research
  process / Panel on Scientific Responsibility and the Conduct of
  Research, Committee on Science, Engineering, and Public Policy,
  National Academy of Sciences, National Academy of Engineering,
  Institute of Medicine.
        p. cm.
    Includes bibliographical references and index.
    ISBN 0-309-04591-6
    1. Research—Moral and ethical aspects. 2. Responsibility.
  I. Title.
  Q180.55.M67C66   1992
  174'.95—dc20   92-10780
                                                               CIP

Printed in the United States of America

First Printing, April 1992
Second Printing, February 1993

The right to search for truth implies also a duty; one must
not conceal any part of what one has recognized to be true.
                                        --Albert Einstein

These words are inscribed on the statue of Albert Einstein that
stands at the front of the National Academy of Sciences building. The
search for truth is the vocation of every scientist, a vocation that
inspires each of us to pursue exciting and controversial ideas, to
engage in spirited exchange with our colleagues and critics, and to
counter customary habits of thinking and analysis with new insights and
observations.

This report, Responsible Science: Ensuring the Integrity of the
Research Process, thoughtfully examines the challenges posed in ensuring
that the search for truth reflects adherence to ethical standards. In
recent years, we have learned, sometimes painfully, that not all
scientists adhere to this obligation. Reports of falsified research
results and plagiarism involving both junior and senior scientists have
stimulated doubts and criticism about the ways in which misconduct in
science is addressed by the research community. Misconduct in science
is now being publicly examined in all of its aspects--how misconduct is
defined, the process by which misconduct is discovered, and procedures
for judging innocence or guilt and assessing penalties. Also being
explored are the appropriate roles of individuals, research
institutions, journals, government research agencies, and the legal
system.

Issues of misconduct and integrity in science present complex
questions. These issues require the sustained attention of all members
of the research community as well as of leaders in the public and
private sector who are concerned with safeguarding the health of
science. In this regard, ensuring the integrity of the research process
is similar to assuring safety in the workplace: it is a process that
requires continued participation from all levels of the entire research
enterprise--the practitioners, the host institutions, the sponsors in
government, and the legislators who provide the funds.

The world of science today is an exciting one, filled with tremendous research opportunities and the ability to contribute to the solution of pressing national needs. In the midst of this excitement, however, it is important to reflect on the values and standards that guide responsible research practices. Three years ago, the Council of the National Academy of Sciences prepared a booklet, <u>On Being a Scientist</u>, to stimulate young researchers to identify and uphold the methods that keep science strong and healthy. <u>Responsible Science</u> is another step toward informing discussions among scientists, and between scientists and the general public, of ethical issues that arise in the contemporary research environment.

Each major institution of American society is now undergoing scrutiny and examination. It is natural for scientists to affirm and protect the traditions and standards that contribute to a healthy and vigorous research system. However, it is also important for scientists to demonstrate the accountability that accompanies public investment in research. This includes setting in place procedures to identify and adjudicate cases of misconduct and supporting measures that will strengthen the integrity of the research system. Cautioning against proposals that may impose counterproductive restraints is important, but not enough.

The report of this panel is the result of intense discussion, analysis, and reflection. It is an important contribution to the national dialogue on integrity in the conduct of research. The broader scientific community knows that ensuring the integrity of the research process is fundamental to the success of science. Scientists must also recognize that it is requisite to the continuing support of science with public funds.

This letter of transmittal conveys the basic sentiments expressed in the report. Ensuring the integrity of the research process is one of the fundamental obligations that accompanies the "right to search for truth."

Frank Press
President
National Academy of Sciences

PATRICIA K. WOOLF, Lecturer, Department of Molecular Biology, Princeton University

KEITH R. YAMAMOTO,* Professor and Vice Chairman, Department of Biochemistry and Biophysics, University of California, San Francisco

Staff

ROSEMARY CHALK, Study Director
BARRY GOLD, Senior Staff Officer
SUSAN MAURIZI, Editor
DAVID H. GUSTON, Research Assistant
MARYANN SHANESY, Administrative Secretary
ELIZABETH BLOUNT, Secretary

---

*Members whose dissent from the panel consensus is expressed in the minority statement following Chapter 8 of the report.

The National Academy of Sciences is a private, nonprofit, self-perpetuating society of distinguished scholars engaged in scientific and engineering research, dedicated to the furtherance of science and technology and to their use for the general welfare. Upon the authority of the charter granted to it by the Congress in 1863, the Academy has a mandate that requires it to advise the federal government on scientific and technical matters. Dr. Frank Press is president of the National Academy of Sciences.

The National Academy of Engineering was established in 1964, under the charter of the National Academy of Sciences, as a parallel organization of outstanding engineers. It is autonomous in its administration and in the selection of its members, sharing with the National Academy of Sciences the responsibility for advising the federal government. The National Academy of Engineering also sponsors engineering programs aimed at meeting national needs, encourages education and research, and recognizes the superior achievements of engineers. Dr. Robert M. White is president of the National Academy of Engineering.

The Institute of Medicine was established in 1970 by the National Academy of Sciences to secure the services of eminent members of appropriate professions in the examination of policy matters pertaining to the health of the public. The Institute acts under the responsibility given to the National Academy of Sciences by its congressional charter to be an adviser to the federal government and, upon its own initiative, to identify issues of medical care, research, and education. Dr. Kenneth I. Shine is president of the Institute of Medicine.

The Committee on Science, Engineering, and Public Policy (COSEPUP) is a joint committee of the National Academy of Sciences, the National Academy of Engineering, and the Institute of Medicine. It includes members of the councils of all three bodies.

Preface

Concerns about integrity in the conduct of research and misconduct in science raise complex issues. Scientists rely on an honor system based on tradition, and on the operation of self-regulating checks and balances to foster responsible research practices. But following a series of highly publicized cases of misconduct in science in the 1980s, the federal government set into motion policies and procedures that now affect every scientist and research institution seeking funding from the Public Health Service and the National Science Foundation. The problems associated with cases of misconduct in science have not yet been resolved. In addition, new concerns have emerged about the methods that are appropriate to ensure integrity in a dynamic, highly decentralized, and diverse research enterprise.

In 1989, the National Academy of Sciences (NAS), the National Academy of Engineering, and the Institute of Medicine (IOM) initiated a major study to examine issues related to scientific responsibility and the conduct of research. The Committee on Science, Engineering, and Public Policy convened a study panel to review factors affecting the integrity of science and the research process as it is carried out in the United States today and to recommend steps for reinforcing responsible research practices. The panel was also asked to review institutional mechanisms that exist for addressing allegations of misconduct in science. Finally, the panel was asked to consider the advantages and disadvantages of formal guidelines for the conduct of research.

The panel included junior and senior scientists from various scientific disciplines, public and private universities, and different regions of the United States; attorneys; research administrators; science editors; a philosopher; a historian; a whistle-blower; and individuals experienced with the formulation of governmental policies on misconduct in science. When our panel of 22 members convened, we knew that this would be a difficult and controversial study.

We did not expect that our discussions would achieve simple solutions or easy explanations for the ethical problems that are apparent in the modern research environment. Panel members had fundamental disagreements about the nature of the problems to be addressed as well as the appropriateness of potential solutions. With two exceptions, panel members achieved consensus in this report. We believe that the ideas, findings, and recommendations that follow provide a foundation for addressing the complex challenge of ensuring the integrity of the research process.

The panel held seven meetings between May 1990 and June 1991. We met with numerous junior and senior scientists, research administrators, government officials, leaders from scientific and educational associations and journals, and congressional representatives. We heard opposing points of view about factors that affect integrity and misconduct in the research environment.

We drew on several Academy studies and reports, including the NAS's *On Being a Scientist* (National Academy Press, Washington, D.C., 1989), the IOM's *The Responsible Conduct of Research in the Health Sciences* (1989), the Government-University-Industry Research Roundtable's *Science and Technology in the Academic Enterprise* (1989), and the National Research Council's *Sharing Research Data* (1985).

Our report consists of two volumes. Volume I includes the findings and recommendations of the study panel. A minority statement, drafted by the two members of the panel who disagreed with the panel consensus, follows Chapter 8. Volume II includes the six working papers of the study panel as well as selected policy statements, developed by various institutions to address issues related to responsible research practices and misconduct in science, that proved useful in the panel's deliberations.

This report recommends specific actions that all scientists, their institutions, and their sponsors can take to preserve and strengthen the integrity of the research process and also to deal with allegations of misconduct. The recommendations provide a blueprint for encouraging and safeguarding the intellectual independence that is essential to doing the best science while also providing for fundamental accountability to those who sponsor and support scientific research.

Edward E. David, Jr., *Chairman*
Panel on Scientific Responsibility and
the Conduct of Research

Acknowledgments

The panel wishes to thank the individuals who provided assistance and information during the course of this study, including Robert Andersen, Defense Nuclear Facilities Safety Board; Michele Applegate, Alcohol, Drug Abuse, and Mental Health Administration; John Bailar, McGill University; Bernard Barber, Columbia University; Michael Barrett, House Committee on Energy and Commerce; Lyle Bivens, U.S. Department of Health and Human Services; Claudia Blair, National Institutes of Health; Erich Bloch, former director, National Science Foundation; James Bower, California Institute of Technology; John Brauman, Stanford University; D. Allan Bromley, Office of Science and Technology Policy; Donald Buzzelli, National Science Foundation; Mary Carter, Agricultural Research Service; Marta Cehelsky, National Science Foundation; Robert Charrow, Crowell and Moring; John Collette, E.I. du Pont de Nemours & Co., Inc.; Tom Devine, Government Accountability Project; Alicia Dustira, Office of Science and Technology Policy; Richard Epstein, University of Chicago; Ned Feder, National Institutes of Health; Nina Fedoroff, Carnegie Institution of Washington; Stephen E. Fienberg, Carnegie Mellon University; Alfred Fishman, University of Pennsylvania; Mark S. Frankel, American Association for the Advancement of Science; Michael Gilman, Cold Spring Harbor Laboratory; D. A. Hendersen, Office of Science and Technology Policy; Charles Herz, National Science Foundation; Roger W. Heyns, the William and Flora Hewlett Foundation; Mark P.

Jacobsen, Covington and Burling; Richard Johns, Johns Hopkins University; Edward Korn, National Heart, Lung and Blood Institute; Donald Langenberg, University of Maryland; and Nathan Lewis, California Institute of Technology.

Also, David Meyer, University of California, Los Angeles; Barbara Mishkin, Hogan and Hartson; Frederick Mosteller, Harvard University; Robert Park, American Physical Society; John Pierce, Stanford University; William Raub, National Institutes of Health; Arnold Relman, *New England Journal of Medicine*; Ellis Rubenstein, *Science* magazine; Paul Russell, Harvard Medical School; Alan Shinn, National Science Foundation; Eleanor Shore, Harvard Medical School; Gregory Simon, House Committee on Science, Space, and Technology; Maxine Singer, Carnegie Institution of Washington; Stephen Smale, University of California, Berkeley; Nicholas Steneck, University of Michigan; Richard Stephens, U.S. Department of Energy; Walter Stewart, National Institutes of Health; Peter Stockton, House Committee on Energy and Commerce; Philip Sunshine, National Science Foundation; Judith Swazey, Acadia Institute; Michael Teitelbaum, the Alfred P. Sloan Foundation; Robert Weinberg, Whitehead Institute; James Wyngaarden, former director, National Institutes of Health; Rosemary Yancik, National Institutes of Health; Larry Zipursky, University of California, Los Angeles; and Diana Zuckerman, House Committee on Government Operations.

Although this report represents the work of the panel members, it would not have been produced without the support of professional staff from the Committee on Science, Engineering, and Public Policy of the National Academy of Sciences, National Academy of Engineering, and Institute of Medicine. Rosemary Chalk, the project's study director, drafted the chapters and refined them on the basis of the panel's discussions and conclusions. Barry Gold, senior staff officer, provided editorial guidance for the report, prepared the material for Chapter 3, and wrote the working paper on congressional activities in Volume II. Dave Guston, research assistant, provided editorial and bibliographic support for the report, prepared contributions for Chapter 2, and wrote the working paper on mentorship in Volume II. Lawrence McCray, executive director of the Committee on Science, Engineering, and Public Policy, provided general guidance and review for the study. The panel is grateful, also, to the secretaries for the project: Maryann Shanesy, Marian Cole, Barbara Candland, and Elizabeth Blount. They prepared manuscripts, arranged travel, and assisted with panel meetings.

Others within the National Academy of Sciences (NAS), Institute of Medicine (IOM), and National Research Council (NRC) who were instrumental in the completion of this study are Frank Press, president of NAS; Samuel O. Thier, former president of IOM; Philip Smith, executive director of the NRC; Enriqueta C. Bond, executive officer of IOM; John Campbell, senior program officer, Government-University-Industry Research Roundtable; Michael A. Stoto, deputy division director, IOM Division of Health Promotion and Disease Prevention; Porter Coggeshall, Report Review Committee; Susan Maurizi, editor, Commission on Physical Sciences, Mathematics, and Applications; and Stephen Mautner, National Academy Press.

Sponsors

This study was undertaken with both public and private sector support. The following agencies of the federal government provided support for the study: the Alcohol, Drug Abuse, and Mental Health Administration, the Department of Agriculture, the Department of Energy, the Department of Health and Human Services, the National Institutes of Health, and the National Science Foundation.

The William and Flora Hewlett Foundation and the Alfred P. Sloan Foundation also awarded grants in support of the study.

Additional support was provided by funds from the National Research Council (NRC) Fund, a pool of private, discretionary, non-federal funds that is used to support a program of Academy-initiated studies of national issues in which science and technology figure significantly. The NRC Fund consists of contributions from a consortium of private foundations including the Carnegie Corporation of New York, the Charles E. Culpeper Foundation, the William and Flora Hewlett Foundation, the John D. and Catherine T. MacArthur Foundation, the Andrew W. Mellon Foundation, the Rockefeller Foundation, and the Alfred P. Sloan Foundation; from the Academy Industry Program, which seeks annual contributions from companies concerned with the health of U.S. science and technology and with public policy issues with technological content; and from the National Academy of Sciences and the National Academy of Engineering endowments.

Contents

xix

Executive Summary

INTRODUCTION

The community of scientists is bound by a set of values, traditions, and standards that embody honesty, integrity, objectivity, and collegiality. These values are reflected in the particular principles and practices characteristic of specific scientific disciplines. The diversity, flexibility, and creativity of the research community—strengths that have contributed to decades of scientific achievement and progress in the United States—also derive from the decentralized character of the research enterprise.

For centuries scientists have relied on each other, on the self-correcting mechanisms intrinsic to the nature of science, and on the traditions of their community to safeguard the integrity of the research process. This approach has been successful largely because of the widespread acknowledgement that science cannot work otherwise, and also because high standards and reputation are important to scientists. Dishonest or untrustworthy individuals become known to their colleagues through various mechanisms, including word of mouth and the inability of other scientists to confirm the work in question. Such irreproducible work is recognized and discredited through the processes of peer review and evaluation that are critical to making professional appointments, accepting work for publication, and awarding research support.

However, the ability of research scientists and their institutions to safeguard the integrity of the research process is now being questioned. Comparatively recent and dramatic increases in the size and influence of the U.S. research enterprise,[1] and in the amounts and patterns of funding, have led to changing social expectations about the accountability of scientists and their institutions for research supported by public funds. In addition, the changing nature of collaborative efforts, the quickening pace and increasing complexity of research endeavors, and the growing emphasis on commercialization of research results have combined to exacerbate stresses that have always been apparent to some extent in scientific research. During the last decade, reports of wrongdoing in science have been accompanied by government oversight and continued scrutiny of the conduct of scientific research. All of these developments have profound implications for the research enterprise's system of internal checks and balances, which evolved in a research environment far removed from the forces of the political process.

The Problem of Misconduct in Science

During the period from March 1989 to March 1991, more than 200 allegations of misconduct in science were recorded by U.S. government offices (NSF, 1990b; Wheeler, 1991).[2] From this number, about 30 cases have resulted so far in confirmed findings of misconduct in science (NSF, 1990b; DHHS, 1991b). Although the possibility of underreporting needs to be considered, these statistics indicate that the reported incidence of misconduct in science is low—compared, for example, to the 26,000 research awards supported annually by the National Institutes of Health (NIH, 1991).

However, any misconduct comes at a high price both for scientists and for the public. Cases of misconduct in science involving fabrication, falsification, and plagiarism breach the trust that allows scientists to build on others' work, as well as eroding the trust that allows policymakers and others to make decisions based on scientific evidence and judgment, especially in instances when definitive studies are not available. The inability or refusal of research institutions to address misconduct-in-science cases can undermine both the integrity of the research process and self-governance by the research community.

Acting to Ensure Integrity in Research

To respond to the need for more visible, explicit mechanisms to ensure integrity in the research process, and to handle allegations of misconduct in science, scientists and their research institutions face three major challenges. One challenge is to develop vigorous approaches to protect and enhance knowledge of scientific traditions and sound research practices and to penalize those who engage in misconduct. A second challenge is to foster responsible research conduct in a period of increasing diversification of funding sources, growing demands on limited research resources, and greater incentives for financial gain in the research environment. A third challenge is to ensure fairness and balance in efforts to establish individual and institutional accountability in scientific research activities, so that frivolous or malicious charges as well as counterproductive regulations are avoided.

PURPOSE AND SCOPE OF THIS STUDY

Charge to the Panel

To address concerns that affect the entire U.S. scientific community, the Committee on Science, Engineering, and Public Policy (COSEPUP) of the National Academy of Sciences, the National Academy of Engineering, and the Institute of Medicine convened the 22-member Panel on Scientific Responsibility and the Conduct of Research. The panel was asked to examine the following issues:

1. What is the state of current knowledge about modern research practices for a range of disciplines, including trends and practices that could affect the integrity of research?
2. What are the advantages and disadvantages of enhanced educational efforts and explicit guidelines for researchers and research institutions? Can the research community itself define and strengthen basic standards for scientists and their institutions?
3. What roles are appropriate for public and private institutions in promoting responsible research practices? What can be learned from institutional experiences with current procedures for handling allegations of misconduct in science?

In addition to outlining approaches to encourage the responsible conduct of scientific research, the panel was also asked to determine whether existing unwritten practices should be expressed as principles to guide the responsible conduct of research. If the panel members judged it advisable, they were encouraged to prepare model guidelines and other materials.

Approach, Scope, and Audience

The panel (1) examined scientific principles and research practices; changes within the contemporary research environment; and the roles of individuals, educational programs, and research guidelines in fostering responsible research practices and (2) considered the incidence and significance of misconduct in science; examined how institutions have handled allegations of misconduct; and also analyzed the complex problems associated with responding to such allegations.

The panel's approach is not intended to diminish the importance of related problems such as conflict of interest and the allocation of indirect costs, but rather to reflect the panel's judgment that integrity in the research process itself and issues arising from misconduct in science deserve critical examination and consideration on their own merits.

Limited availability to date of evaluated data and the fact that the panel often had to rely on its own informed judgment require that this report be viewed as part of a comprehensive dialogue on and examination of integrity in the research process. The panel emphasizes that this report is addressed to *all* members of the scientific community, regardless of their institutional affiliation.

Defining Terms—Articulating a Framework for Fostering Responsible Research Conduct

The panel defined the term "integrity of the research process" as the adherence by scientists and their institutions to honest and verifiable methods in proposing, performing, evaluating, and reporting research activities.

To provide policy guidance for scientists, research institutions, and government research agencies concerned about ensuring the integrity of the research process as well as addressing misconduct in science, the panel developed a framework that delineates three categories of behaviors in the research environment that require attention. These categories are (1) misconduct in science, (2) questionable research practices, and (3) other misconduct.

Unethical actions of all types are intolerable, and appropriate actions by the research community to address such problems are essential. But the panel believes that there are risks inherent in developing institutional policies, procedures, and programs that treat all of these behaviors without distinction. Inappropriate actions by government and institutional officials can create an atmosphere that disturbs effective methods of self-regulation and harms pioneering research activities.

In developing its framework of definitions, the panel adopted an approach that evaluates how seriously the various behaviors compromise the integrity of the research process.

Misconduct in Science

Misconduct in science is defined as fabrication, falsification, or plagiarism, in proposing, performing, or reporting research. Misconduct in science does not include errors of judgment; errors in the recording, selection, or analysis of data; differences in opinions involving the interpretation of data; or misconduct unrelated to the research process.

Fabrication is making up data or results, falsification is changing data or results, and plagiarism is using the ideas or words of another person without giving appropriate credit.

By proposing this precise definition of misconduct in science, the panel is in unanimous agreement that the core of the definition of misconduct in science should consist of fabrication, falsification, and plagiarism. The panel unanimously rejects ambiguous language such as the category "other serious deviations from accepted research practices" currently included in regulatory definitions adopted by the Public Health Service and the National Science Foundation (DHHS, 1989a; NSF, 1991b). In particular, the panel wishes to discourage the possibility that a misconduct complaint could be lodged against scientists based solely on their use of novel or unorthodox research methods. The use of ambiguous terms in regulatory definitions invites exactly such an overexpansive interpretation.

In rejecting the "other serious deviations" category, the panel considered whether a different measure of flexibility should be included in its proposed definition of misconduct in science, so as to allow the imposition of sanctions for conduct similar in character to fabrication, falsification, and plagiarism.

Some panel members believe that the definition should also encompass other actions that directly damage the integrity of the research process and that are undertaken with the intent to deceive.

Questionable Research Practices

Questionable research practices are actions that violate traditional values of the research enterprise and that may be detrimental to the research process. However, there is at present neither

broad agreement as to the seriousness of these actions nor any consensus on standards for behavior in such matters. Questionable research practices do not directly damage the integrity of the research process and thus do not meet the panel's criteria for inclusion in the definition of misconduct in science. However, they deserve attention because they can erode confidence in the integrity of the research process, violate traditions associated with science, affect scientific conclusions, waste time and resources, and weaken the education of new scientists.

Questionable research practices include activities such as the following:

- Failing to retain significant research data for a reasonable period;
- Maintaining inadequate research records, especially for results that are published or are relied on by others;
- Conferring or requesting authorship on the basis of a specialized service or contribution that is not significantly related to the research reported in the paper;[3]
- Refusing to give peers reasonable access to unique research materials or data that support published papers;
- Using inappropriate statistical or other methods of measurement to enhance the significance of research findings;[4]
- Inadequately supervising research subordinates or exploiting them; and
- Misrepresenting speculations as fact or releasing preliminary research results, especially in the public media, without providing sufficient data to allow peers to judge the validity of the results or to reproduce the experiments.

The panel wishes to make a clear demarcation between misconduct in science and questionable research practices—the two categories are not equivalent, and they require distinct types of responses by the research community and research institutions.

Other Misconduct

Certain forms of unacceptable behavior are clearly not unique to the conduct of science, although they may occur in a laboratory or research environment. Such behaviors, which are subject to generally applicable legal and social penalties, include actions such as sexual and other forms of harassment of individuals; misuse of funds; gross negligence by persons in their professional activities; vandalism, including tampering with research experiments or instrumentation; and violations of government research regulations, such as those dealing with radioactive materials, recombinant DNA research, and the use of human or animal subjects. Industry-university relationships, and the resultant possibility of conflicts of interest, also raise issues that require special attention.

Recognized legal and institutional procedures should be in place to address complaints and to discourage behavior involving forms of misconduct that are not unique to the research process. The panel concluded that such behaviors require serious attention but lie outside the scope of the charge for this study.

On some occasions, however, certain forms of "other misconduct" are directly associated with misconduct in science. Among these are cover-ups of misconduct in science, reprisals against whistle-blowers, malicious allegations of misconduct in science, and violations of due process protections in handling complaints of misconduct in science. These forms of other misconduct may require action and special administrative procedures.

FINDINGS AND CONCLUSIONS

Scientists and Research Institutions

Because scientists and the achievements of science have earned the respect of society at large, the behavior of scientists must accord not only with the expectations of scientific colleagues, but also with those of a larger community. As science becomes more closely linked to economic and political objectives, the processes by which scientists formulate and adhere to responsible research practices will be subject to increasing public scrutiny. This is one reason for scientists and research institutions to clarify and strengthen the methods by which they foster responsible research practices.

Accordingly, the panel emphasizes the following conclusions:

• The panel believes that the existing self-regulatory system in science is sound. But modifications are necessary to foster integrity in a changing research environment, to handle cases of misconduct in science, and to discourage questionable research practices.

• Individual scientists have a fundamental responsibility to ensure that their results are reproducible, that their research is reported thoroughly enough so that results are reproducible, and that significant errors are corrected when they are recognized. Editors of scientific journals share these last two responsibilities.

• Research mentors, laboratory directors, department heads, and senior faculty are responsible for defining, explaining, exemplifying, and requiring adherence to the value systems of their institutions.

• Administrative officials within the research institution also bear responsibility for ensuring that good scientific practices are observed in units of appropriate jurisdiction and that balanced reward systems appropriately recognize research quality, integrity, teaching, and mentorship.

The Changing Research Enterprise

The academic research community, governed by traditions derived from an earlier model of a community of independent scholars who participated equally in academic governance, is challenged by the complexity of today's issues and of the environment in which research is conducted. Still, basic research continues to flourish, and faculty, postdoctoral fellows, and graduate students continue to contribute extraordinary research capability to science.

In reviewing changes within the scientific research enterprise, the panel reached the following conclusions:

• Scientific research is part of a larger and more complicated enterprise today, creating a greater need for individual and institutional attention to matters that affect the integrity of the research process. Scientists themselves and research institutions will be expected to play a more active role in ensuring that the activities performed by researchers are within the governance mechanisms of their institutions.

• The growth and diversity of modern research call for institutions to accept explicit responsibility for fostering the integrity of the research process and for handling allegations of misconduct. In recognizing that their faculty and research staff are responsible for maintaining the integrity of the research process, institutions should retain and accept certain explicit obligations. Principal among these is providing a research environment that fosters honesty, integrity, and a sense of community. Research institutions should also recognize the risks that are inherent in self-regulation and strive to involve outside parties, when appropriate, in investigating or evaluating the conduct of their own members.

• The increased size, specialization, and diversity of research groups, and other changes in the social relationships of their members, have stimulated personal conflicts and misunderstandings, including disputes about fairness and allocation of credit. These disputes may be prevented by positive efforts to foster responsible research practices and by taking preemptive actions to promote a harmonious and productive workplace. Frank discussions, both formal and informal, possibly aided by outside mediators, are additional tools to use in addressing these disputes.

• The issues associated with conflict of interest in the academic research environment are sufficiently problematic that they deserve thorough study and analysis by major academic and scientific organizations, including the National Academy of Sciences.

• The research environment is stressful and yet conducive to the remarkable productivity of researchers. The rewards for successful research are greater now than in the past, but today's rapid pace of development may undermine critical internal checks and balances and may increase opportunities for misrepresentation or distortion of research results.

Misconduct in Science—Incidence and Significance

The panel found that existing data are inadequate to draw accurate conclusions about the incidence of misconduct in science or of questionable research practices. **The panel points out that the number of confirmed cases of misconduct in science is low compared to the level of research activity in the United States. However, as with all forms of misconduct, underreporting may be significant; federal agencies have only recently imposed procedural and reporting requirements that may yield larger numbers of reported cases. Any misconduct comes at a price to scientists, their research institutions, and society. Thus every case of misconduct in science is serious and requires attention.**

Handling Allegations of Misconduct in Science— Institutional Responses and Experience

University-Government Approaches

Government agencies, congressional oversight committees, and academic institutions generally agree that *the primary responsibility for handling complaints of misconduct in science rests with the research organization.* However, the development and implementation of policies and procedures for handling misconduct in science have been problematic. Some universities, particularly small research institutions, are not prepared to accept responsibility for pursuing allegations of misconduct in science.[5] It is difficult for any institution to investigate members of its own community, especially individuals who hold positions of high esteem. In addition, some research institutions and government agencies have made mistakes in investigations of complex cases, such as appointing to investigatory panels members who have personal or professional ties to the individuals who have been accused of misconduct in science. All these factors foster a perception that research institutions are not dealing effectively with misconduct in science,[6] prompting criticism of the speed, rigor, honesty, fairness, and openness of their response mechanisms.

Many universities have now established policies and procedures for handling allegations of misconduct in science, and some research institutions have acquired valuable experience in implementing these procedures to deal with cases of misconduct. However, the legal and procedural issues associated with misconduct-in-science investigations are extraordinarily complex, and there is little case law in the public record to guide and inform analysis of these issues.

The panel believes that, in general, the current and evolving system of government and institutional relationships requires more experience and adjustments before specific policy or procedural changes can be recommended. Research institutions need to clarify their own approaches and judgments on these issues before any general consensus can be reached on procedural matters.

Part of the difficulty in developing vigorous and effective institutional responses to incidents or allegations of misconduct in science arises from variation in and disagreement about essential elements of fairness, completeness, and objectivity that should characterize investigations. Effective responses are impeded also by recurring patterns of denial by some institutional officials and faculty members who believe that misconduct in science is not a serious matter. The pressures of conducting an objective investigation of complaints involving respected or prestigious scientists cannot be underestimated. Strong and informed leadership is needed to clarify procedural matters and to ensure that allegations or apparent incidents of misconduct in science are not ignored or covered up.

Need for Explicit Procedural Elements

Institutional policies and procedures should include a common entry point for handling complaints from the outset; clear procedures are necessary for determining which type of alleged offenses will be reviewed by administrative staff or faculty. A sequence of steps to achieve resolution of significant disputes is required. All of these steps require clear separations between each of the following groups: the affected parties, those who are judging the seriousness of the complaint and formulating the evidentiary base to substantiate charges, and those who must adjudicate penalties based on charges of misconduct in science.

The panel believes that institutional procedures should define explicit and clear criteria that are to be used in determining when a misconduct inquiry should proceed to a more formal investigation. The panel concludes that administrative officials and faculty have a responsibility to inform all members of their institution, especially junior personnel, of existing channels for handling complaints about misconduct in science or other misconduct.

Current Situation

The panel is aware of the inherent difficulty posed by asking research institutions to investigate allegations of misconduct in science that involve their own members. Internal investigations must demonstrate a fundamental commitment to independence and objectivity to ensure their credibility and success, and may be enhanced by the participation of members from outside the affected organization. The objectivity of misconduct-in-science investigations also relies heavily on the credibility of the process used to arrive at findings and recommendations. To maintain the privilege of self-regulation, research institutions must exercise vigilance and diligence in examining the conduct of their own members.

Balancing Accountability and the
Need for Intellectual Freedom

In the wake of procedural and policy reforms in response to incidents of misconduct in science, representatives from the academic and scientific community have raised concerns about the long-term or unintended effects that might result from institutional or governmental intrusions into the research environment.[7] Aggressive efforts to control research practices, if carried to an extreme, can damage the research enterprise. Balance is required. Inflexible rules or requirements can increase the time and effort necessary to conduct research, can discourage creative individuals from pursuing research careers, can decrease innovation, and can in some instances make the research process impossible. Governmental or regulatory efforts to define "correct" research conduct or analytical practices can do fundamental harm to research activities if such efforts encourage orthodoxy and rigidity and inhibit novel or creative research practices.

However, the panel concludes that allegations and incidents of misconduct in science require a vigorous institutional response and that the methods used by research institutions and government to address allegations of misconduct in science need improvement. Research institutions sometimes require advice or assistance in addressing allegations of misconduct in science because of the complexities of these cases or because their faculty or administrators are reluctant to address in a systematic manner complaints or suspicions about possible misconduct in science. Research institutions have not developed mechanisms for broad exchange of information and experience in resolving difficult cases and consequently lack opportunities for learning from each other.

Steps to Encourage Responsible Research Practices

In considering different approaches to dealing with questionable research practices, the panel concluded that questionable practices are best discouraged through peer review and the system of appointments, evaluations, and other rewards in the research environment as well as educational programs that emphasize responsible behavior in the research environment. Such approaches build on the strengths of self-regulation, rely on those who are most knowledgeable about the intricacies of the scientific process to maintain the quality of the research environment, and preserve the diverse disciplinary traditions that are essential to responsible scientific conduct. By encouraging the development of educational programs that emphasize responsible research behavior, the panel seeks to foster more deliberate and informed communication, discussion, criticism, and reflection of the basic values that guide scientific practices and judgments.

In considering the advantages and disadvantages of guidelines for research conduct, the panel concluded that although the process of formulating guidelines may be extremely valuable for those who participate, guidelines that are relevant and appropriate to research may vary considerably depending on the research field, the nature of the work, and other factors. To be effective, guidelines must be incorporated into the process of research and education and become an operational part of day-to-day activities. If faculty desire to develop guidelines for the conduct of research, such policies should be formulated by those who will be directly affected and should be adapted to specific research fields and protocols.

Institutional guidelines are likely to be less effective than ones formulated at the group or laboratory level. However, research institutions may wish to adopt an overarching set of general principles for their members to provide a common frame of reference. The panel recognizes that the formulation of written guidelines is an exacting task that requires substantial time and effort.

The panel concluded that subjects such as data management, publication practices, authorship, peer review, and training and supervision should be considered in any efforts aimed at developing educational discussions or guidelines for the responsible conduct of scientific research. This set of subjects suggests particular topics and examples of "best scientific practice" that should be considered in formulating statements on research conduct.

RECOMMENDATIONS

Ensuring the integrity of the research process requires that scientists and research institutions give systematic attention to the fundamental values, principles, and traditions that foster responsible research conduct. In considering factors that may affect integrity and misconduct in science, the panel formulated twelve recommendations to strengthen the research enterprise and to clarify the nature of the responsibilities of scientists, research institutions, and government agencies in this area.

Acting to Define and Strengthen Basic Principles and Practices

Recommendation One

Individual scientists in cooperation with officials of research institutions should accept formal responsibility for ensuring the integrity of the research process. They should foster an environment, a reward system, and a training process that encourage responsible research practices.

Recommendation Two

Scientists and research institutions should integrate into their curricula educational programs that foster faculty and student awareness of concerns related to the integrity of the research process.

Recommendation Three

Adoption of formal guidelines for the conduct of research can provide a valuable opportunity for faculty and research institutions to clarify the nature of responsible practices, but adopting guidelines should be an option, not a requirement, for research institutions.

Dealing with Misconduct—Institutional Roles

Recommendation Four

Research institutions and government agencies should adopt a common framework of definitions, distinguishing among misconduct in science, questionable research practices, and other forms of misconduct. They should adopt a single consistent definition of misconduct in science that is based on fabrication, falsification, and plagiarism. Accordingly, federal agencies should review their definitions of misconduct in science to remove ambiguous categories such as "other serious deviations from accepted research practices."

Recommendation Five

Government agencies should adopt common policies and procedures for handling allegations of misconduct in science. The Office of Science and Technology Policy (OSTP) should lead the effort to establish government-wide definitions and procedures. OSTP should consider adopting the definition of misconduct in science proposed in this report and use this definition in formulating government-wide model policies.

Recommendation Six

Research institutions and government research agencies should have policies and procedures that ensure appropriate and prompt responses to allegations of misconduct in science. Research institutions should foster effective and appropriate methods for detecting and handling incidents of misconduct in science and should strengthen the implementation of misconduct-in-science policies and procedures that incorporate fundamental elements of due process.

Recommendation Seven

Scientists and their institutions should act to discourage questionable research practices through a broad range of formal and informal methods in the research environment. They should also accept responsibility for determining which questionable research practices are serious enough to warrant institutional penalties. But the methods used by individual scientists and research institutions to address questionable research practices should be distinct from those for handling misconduct in science or other misconduct.

Recommendation Eight

Research institutions should have policies and procedures to address other misconduct—such as theft, harassment, or vandalism—that may occur in the research environment. Where procedures for handling complaints about other misconduct do not exist, allegations should be examined according to the same administrative mechanisms as those designed to address misconduct in science, although the procedural pathways for responding to other misconduct and misconduct in science may differ.

Recommendation Nine

Government research agencies should clarify their roles in addressing misconduct in science, questionable research practices, and other misconduct. Although government agencies have specific regulatory responsibilities in handling the categories of misconduct in science and other misconduct, their role in addressing questionable research practices should be designed to support the efforts of scientists and research institutions to discourage such practices through the processes of education and peer review.

Taking Additional Steps

Recommendation Ten

An independent Scientific Integrity Advisory Board should be created by the scientific community and research institutions to exercise leadership in addressing ethical issues in research conduct; in framing model policies and procedures to address misconduct in science and other misconduct; to collect and analyze data on episodes of misconduct in the research environment; to provide periodic assessments of the adequacy of public and private systems that have been developed to handle misconduct in science cases; and to facilitate the exchange of information about and experience with policies and procedures governing the handling of allegations of misconduct in science.

Recommendation Eleven

The important role that individual scientists can play in disclosing incidents of misconduct in science should be acknowledged. Individuals who, in good conscience, report suspected misconduct in science deserve support and protection. Their efforts, as well as the efforts of those who participate in misconduct proceedings, can be invaluable in preserving the integrity of the research process. When necessary, serious and considered whistle-blowing is an act of courage that should be supported by the entire research community.

Recommendation Twelve

Scientific societies and scientific journals should continue to provide and expand resources and forums to foster responsible research practices and to address misconduct in science and questionable research practices.

NOTES

1. Government funding for U.S. basic research increased in current dollars from $5.4 billion in FY 1982 to an estimated $12.5 billion in FY 1991. See p. 53 in American Association for the Advancement of Science (1991a). Academic research investigators are also increasingly supported by nonfederal funds provided by a diverse mix of industrial sponsors, state, and local funds, foundations, and intramural support. For example, the industrial share of academic R&D funding grew from 3.9 percent in 1980 to an estimated 6.6 percent in 1989. Some specialized academic research centers now receive over 20 percent of their funding from industry. See p. 106 in National Science Board (1989).

2. The term "allegation" here refers to complaints of misconduct in science that have resulted in a government case file. An analysis of these allegations is provided in Chapter 4. As of December 1991, about half of these allegations had been resolved.

3. It is possible that some extreme cases of noncontributing authorship may be regarded as misconduct in science because they constitute a form of falsification. These would include only cases in which an individual who has made *no* identifiable contribution to a research paper is named, or seeks to be named, as a co-author.

4. See Bailar (1986).

5. See, for example, the discussion in the DHHS's OIG report (DHHS, 1989d), which notes that although all "large grantee institutions considered [misconduct] investigations their responsibility, . . . only 54 percent of the small institutions shared this view, and most of these institutions would support a more active NIH role in investigating allegations" (p. 11).

6. See the statement by Rep. John Dingell in U.S. Congress (1989b): "The apparent unwillingness on the part of the scientific community to deal promptly and effectively with allegations of misconduct is unfair to both the accuser and to the accused" (p. 1). See also Weiss (1991b) and the commentary in Dong (1991).

7. See, for example, testimony by academic officials and scientists in hearings convened by the House Committee on Science, Space, and Technology (U.S. Congress, 1990b).

1

Introduction

THE U.S. RESEARCH ENTERPRISE

The Traditions of Science

Scientific research is grounded in values such as integrity, honesty, trust, curiosity, and respect for intellectual achievement.[1] The expression of these values in the diverse styles and approaches of the various scientific disciplines has contributed directly to the discovery of knowledge and thus to the achievements of the U.S. scientific research enterprise. Basic to the honor system that binds the community of scientists is truthfulness, both as a moral imperative and as a fundamental operational principle in the scientific research process.[2] Ideally, it is the challenge of gaining a measure of truth that motivates scientists to formulate, test, and revise their hypotheses in ways that minimize errors.

Scientific achievement as well as human welfare, which is affected increasingly by the work of scientists, depend on the integrity of the research process. By *integrity of the research process*, the panel means the adherence by scientists and their institutions to honest and verifiable methods in proposing, performing, evaluating, and reporting research activities. The research process includes the construction of hypotheses; the development of experimental and theoretical paradigms; the collection, analysis, and handling of data; the generation of new ideas, findings, and theories through experimentation and analysis; timely communication and publication; refinement of results through replication and extension of the original work; peer review; and the training and supervision of associates and students. The traditions of skepticism, openness, sharing, and disclosure that are associated with the research process not only provide a means of identifying theoretical or experimental errors that occur inevitably in science, but also imply an obligation to maintain the integrity of the research process. Errors are often corrected by later research, stimulated by the skepticism of other scientists. Error, however, is distinct from actions that directly compromise the integrity of the research process.

Scientists have relied on each other and the traditions of their community for centuries to safeguard the integrity of the research process. This approach has been successful largely because of the widespread acknowledgement that science cannot work otherwise, and also because high standards and reputation are important to scientists. Dishonest or untrustworthy individuals become known to their colleagues through various mechanisms, including word of mouth and the inability of other scientists to confirm the work in question. Such irreproducible work is recognized and discredited through the processes of peer review and evaluation that are critical to making professional appointments, accepting work for publication, and awarding research support.

Changing Circumstances and Expectations

The U.S. scientific community has maintained a high degree of autonomy and self-governance during a period of remarkable successes. But the ability of research scientists and their institutions to safeguard the integrity of the research process is now being questioned as a result of several significant and comparatively recent developments.[3]

Among these developments are the dramatic increases in the size of the U.S. research enterprise and in the amounts and patterns of funding.[4] These increases have come in response to the many notable contributions of scientists, engineers, and health professionals, emerging research opportunities, and public demands for solutions to such complex problems as protecting the environment and ensuring economic well-being. Also apparent are pressures related to the quickening pace and use of new developments in science—research results in some areas can rapidly influence public policy, health care services, and the commercial value of new products.

By many measures, the U.S. research system has remained notably creative and productive in this changing environment. In addition to advances in knowledge about the fundamental processes of biological, physical, and social systems, a major achievement has been the creation of a generation of well-trained research investigators. The very success of these and other scientific achievements has contributed to an expanding research enterprise.

But the new dimensions of the scientific enterprise do not come without stress (Hackett, 1990; OTA, 1991). The growth in the size and specialization of research teams in some fields has strained the capacity of individual scientists to maintain the degree of personal involvement and familiarity with their colleagues' and subordinates' efforts that characterized earlier work. In the words of one scientist: "It is increasingly difficult for a scientist to master, let alone know in detail, the reliability of every phase of a large, multiple-author work. Thus, the individual scientist depends upon the integrity and competence of colleagues" (Hoshiko, 1991, p. 11).

Individual and institutional efforts to manage and resolve internal stresses in the research enterprise are increasingly apparent (Hackett, 1990; OTA, 1991). Disputes have arisen among scientists over allocation of credit and recognition of intellectual property rights. Schisms have increased between scientific investigators and research administrators, sponsors, and funders over issues such as research budgets, cost accounting for research activities, the appropriate allocation of time between teaching and research responsibilities, and the level of oversight for research activities.

As a result, scientists are calling attention to factors in the research environment itself that have been identified as negative.[5] But questions about how to safeguard the integrity of an enterprise that is central to contemporary American life are of concern to more than scientists alone. The self-regulatory system in science, which has evolved over the centuries to foster creativity and scientific achievement, may need to evolve further to meet the demands for public accountability that accompany government, foundation, and industrial support of scientific research.[6]

The Problem of Misconduct in Science

In the 1980s, newspaper and magazine accounts brought to the attention of the scientific community, the public at large, and the federal government several instances of scientists who reported measurements they never made, altered research results, or plagiarized the work of others.[7] In many cases, the responses of the institutions where these actions had occurred seemed slow, inadequate, and confused.[8] Some research institutions and government agencies set up investigations that appeared to be biased or failed to disclose incidents of misconduct. Such events raised additional questions about the integrity of the research process and about the traditional self-governance of the scientific research community. Continuing concern has been fueled by anecdotal evidence emerging from press reports, congressional hearings, or institutional actions taken in specific cases.

During the period from March 1989 to March 1991, more than 200 allegations of misconduct in science were recorded by U.S. government offices (NSF, 1990b; Wheeler, 1991).[9] From this number, about 30 cases have resulted so far in confirmed findings of misconduct in science (NSF, 1990b; DHHS, 1991b). Although the possibility of underreporting needs to be considered, these statistics indicate that the reported incidence of misconduct in science is low—compared, for example, to the 26,000 research awards supported annually by the National Institutes of Health (NIH, 1991).

But any misconduct comes at a high price both for scientists and for the public, and the possibility of underreporting needs to be considered. Cases of misconduct in science involving fabrication, falsification, and plagiarism breach the trust that allows scientists to build on others' work, as well as eroding the trust that allows policymakers and others to make decisions based on scientific and objective evidence. The inability or refusal of research institutions to address such cases can undermine both the integrity of the research process and self-governance by the research community.

ENSURING INTEGRITY IN THE RESEARCH PROCESS

Acknowledging the Range of Concerns

Scientists are deeply troubled by reports of misconduct in science. At the same time, they are concerned that institutional and government bureaucracies designed to uncover or respond to allegations of misconduct in science may damage the vitality and productivity of U.S. scientific research. The creative processes of scientific judgment, experimentation, and error-correction that are intrinsic to the development of new scientific knowledge require a flexible and adaptable environment. In a time of expanding research opportunities and competitive funding pressures, many scientists also fear that significant time, and possibly resources, could be diverted from research endeavors and used instead to satisfy administrative controls derived from political imperatives.

In 1985, the U.S. Congress passed legislation that requires each institution receiving funds from the Public Health Service to develop an "administrative process to review reports of scientific fraud in connection with biomedical or behavioral research" sponsored by the institution.[10] Later, the Public Health Service and the National Science Foundation each adopted regulations designed to address allegations of misconduct in science (DHHS, 1989a; NSF, 1987, 1991b).

Such legislative and regulatory decisions concerned with the integrity of the research process and misconduct in science mark the beginning of a new set of relationships between the scientific community and the federal government. In contrast to an earlier period characterized by unwritten agreements and personal trust, current policy discussions about integrity and misconduct in science raise difficult questions about the roles of individual scientists, their laboratories, research institutions, and government in providing oversight of the research enterprise to ensure that science is conducted in an honest and responsible manner.

One observer has aptly summarized some of the basic concerns (Menninger, 1990):

> The appearance of federal policy [in an area] once mainly confined to academic and scientific circles must be taken as a significant matter reflecting a heightened public perception of scientific research as a determinant of the national future. This results not just from scientific research's role as producer of new knowledge, but from its expanding need for sizeable sums of money, its impact on the country's economic prospects and quality of life, and its generation of painfully complex ethical dilemmas. These factors have unequivocally moved the research enterprise out of the isolation of the laboratory and onto the public stage where a context of motives and expectations prevails that scientists may find incongruous with their vocation, but which they ignore at their peril.

Taking Constructive Action

Expectations within and outside of research institutions have generated desires for more visible, explicit mechanisms to handle allegations of misconduct in science and to assure integrity in the research process. One challenge is to develop vigorous approaches to protect and enhance scientific traditions and sound research practices and to penalize those who engage in misconduct. A second challenge is to foster responsible research conduct in a period of increasing diversification of funding sources, growing demands on limited research resources, and greater incentives for financial gain in the research environment. A third challenge is to ensure fairness and balance in efforts to establish individual and institutional accountability in publicly supported research activities, so that frivolous or malicious charges as well as counterproductive regulations are avoided.

PURPOSE AND SCOPE OF THIS STUDY

Charge to the Panel

The Committee on Science, Engineering, and Public Policy (COSEPUP) of the National Academy of Sciences, the National Academy of Engineering, and the Institute of Medicine sought to address these issues by convening the 22-member Panel on Scientific Responsibility and the Conduct of Research. The panel was asked to examine the following issues:

1. What is the state of current knowledge about modern research practices for a range of disciplines, including trends and practices that could affect the integrity of research?

2. What are the advantages and disadvantages of enhanced educational efforts and explicit guidelines for researchers and research institutions? Can the research community itself define and strengthen basic standards for scientists and their institutions?

3. What roles are appropriate for public and private institutions in promoting responsible research practices? What can be learned from institutional experiences with current procedures for handling allegations of misconduct in science?

In addition to outlining approaches to encourage the responsible conduct of scientific research, the panel was also asked to determine whether existing unwritten practices should be expressed as principles to guide the responsible conduct of research. If the panel members judged it advisable, they were encouraged to prepare model guidelines and other materials.

Approach, Audience, Content

In responding to its charge, the panel chose a two-part approach intended to produce a report that would speak to all members of the U.S. scientific research community. First, the panel examined factors fundamental to the integrity of the research process, including scientific principles and research practices; changes within the contemporary research environment; and the roles of individuals, educational programs, and research guidelines in fostering responsible research practices. Second, the panel considered the incidence and significance of misconduct in science and also examined institutional approaches to handling allegations of misconduct, analyzing in addition the complex problems associated with responding to such allegations.

The panel chose this approach to emphasize positive steps that might be taken to assure the integrity of the research process in the current environment. Although many organizations are absorbed with responding to the problem of misconduct in science, institutional experience with recently adopted regulatory requirements is very new, and there is not yet a clear consensus about procedural approaches that may be necessary to address allegations of misconduct.[11] The panel did not attempt to resolve all of these matters in this report. At the same time, its selected approach is not intended to diminish the importance of related problems such as conflict of interest, the allocation of indirect costs, or federal support for scientific research, but rather to reflect the panel's judgment that integrity in the research process itself and issues arising from misconduct in science deserve critical examination and consideration on their own merits.

Although this report addresses concerns that affect the entire U.S. scientific community, the members of the panel were obliged to generalize from their own particular specialized expertise and experience. Unfortunately, it was not possible to develop a detailed description of the diverse styles and approaches of the various scientific disciplines, a description that might have conveyed the richness, spirit, and disciplinary differences that characterize U.S. science. The panel recognizes this limitation but believes that a general approach will guide specific interpretations and applications. This report should therefore be viewed as part of a comprehensive dialogue on and examination of integrity in the research process.

Finally, the panel emphasizes that this report speaks to *all* members of the scientific community, regardless of their institutional affiliation, whose research results become part of the scientific process. Although this report is addressed principally to an academic audience, the panel believes that the discussions, findings, and recommendations also have relevance for nonacademic research groups, including those in industry, and particularly those engaged in clinical trials and drug toxicology studies, as well as others whose members report research results at scientific meetings and publish in journals. Officials at research institutions also are responsible for taking educational, preventive, and remedial approaches to dealing with scientific conduct issues. All who participate in the research enterprise share responsibility for the integrity of the research process.

METHODS, DEFINITIONS, AND BASIC ASSUMPTIONS

Evaluating Available Data

The panel sought to develop a report that would address conflicting perspectives and priorities basic to enhancing integrity in the research process. An examination of empirical studies on research behaviors yielded few significant insights.[12]

The panel also concluded that existing social studies of the U.S. scientific research enterprise are not adequate to support conclusions about the relative effectiveness of various alternatives for fostering the integrity of the research process. For example, the value of formal and informal educational approaches in fostering responsible research practices has, to the panel's knowledge, not been systematically addressed. And although some research institutions in recent years have adopted formal guidelines designed to foster responsible practices, the experience with research guidelines is limited.[13]

The panel also found barriers to obtaining data on specific incidents of misconduct. Confidential institutional reports are not available if misconduct cases are under appeal or are subject to litigation, if the institutions have negotiated private settlements with the subjects of misconduct complaints, if there are findings of no misconduct, or if the misconduct has been judged to be not significant enough to warrant penalties. Those involved in handling or evaluating misconduct cases are usually not at liberty to discuss their findings. Those who have been parties at interest in misconduct cases may have a biased view of specific actions. An increasing amount of litigation in misconduct cases has further complicated the collection and analysis of primary data.

Thus many of the panel's findings and recommendations are derived from informed judgments based on discussions with persons knowledgeable about the research process and about factors that affect the contemporary research environment. The panel also met with individuals who have both knowledge of and a broad range of perspectives on the significance of the reported cases of misconduct in science. The panel's overall outlook and opinions are based on general ethical principles that are well accepted by scientists and by society.

Defining Terms—Articulating a Framework for
Fostering Responsible Research Conduct

In the opening paragraphs of this chapter, the panel defined the term "integrity of the research process" as the adherence by scientists and their institutions to honest and verifiable methods in proposing, performing, evaluating, and reporting research activities. This term is sometimes thought to be synonymous with "integrity of science," but the terms of reference are different.[14] Science is not only a body of information, composed of current knowledge, theories, and observations, but also the process by which this body of knowledge is developed. Furthermore, the scientific process is a social enterprise that involves individuals and institutions engaged in developing, certifying, and communicating research results. Throughout this report the panel focuses on the integrity of the research process as defined above.

Misconduct in science is commonly referred to as fraud.[15] But most legal interpretations of the term "fraud" require evidence not only of intentional deception but also of injury or damage to victims. Proof of fraud in common law requires documentation of damage incurred by victims who relied on fabricated or falsified research results. Because this evidentiary standard seemed poorly suited to the methods of scientific research, "misconduct in science" has become the common term of reference in both institutional and regulatory policy definitions.

However, "misconduct in science" as commonly used is an amorphous term, often covering a spectrum of both significant and trivial forms of misbehavior by scientists. The absence of a clear, explicit definition that focuses on actions highly detrimental to the integrity of the research process has impeded the development of effective institutional oversight and government policies and procedures designed to respond to such actions. Varying definitions of misconduct in science have also impeded comparison of the results of survey studies. If, for example, survey respondents apply the term "misconduct in science" to a broad range of behaviors that extend beyond legal or institutional definitions, their responses weaken the significance of reported survey results.

In order to provide policy guidance for scientists, research institutions, and government research agencies concerned about ensuring the integrity of the research process as well as addressing misconduct in science, the panel developed a framework that delineates three categories of behaviors in the research environment that require attention. These categories are (1) misconduct in science, (2) questionable research practices, and (3) other misconduct.

The panel seeks to accomplish several goals by proposing these three categories. Foremost is a precise definition of misconduct in science aimed at identifying behaviors that scientists agree seriously damage the integrity of the research process. For example, although using inadequate training methods or refusing to share research data or reagents are not desirable, such actions generally are regarded as behaviors that are not comparable to the fabrication of research data. In the same manner, sexual harassment and financial mismanagement are illegal behaviors regardless of whether scientists are involved, but these actions are different from misconduct in science because they do not compromise, in a direct manner, the integrity of the research process.

Unethical actions of all types are intolerable, and appropriate actions by the research community to address such problems are essential. But the panel believes that there are risks inherent in developing institutional policies, procedures, and programs that treat all of these behaviors without distinction. Inappropriate actions by government and institutional officials can create an atmosphere that disturbs effective methods of self-regulation and harms pioneering research activities. In particular, many scientists are concerned that the term "misconduct in science," which has been construed as including "serious deviations from accepted practices" (as currently defined in government regulations), could be defined in such a way that it could be applied inappropriately to the activities of honest scientists engaged in creative research efforts.

The panel recognizes that this framework may not satisfy all scientists, lawyers, or policymakers. Its primary purpose is to advance the quality of policy and educational discussions about distinctions between different kinds of troubling behavior within the research environment, and to allow scientists, institutional officers, and public officials to focus their attention and their efforts toward prevention on substantive issues rather than discrepancies in terminology. Thus the framework of definitions proposed in this report should be viewed as a tool for use in a sustained effort by the research community to strengthen the integrity of the research process, to promote responsible research conduct, and to clarify appropriate methods to address instances of misconduct in science. The three categories will need to be refined through continued dialogue, criticism, and experience.

In developing its framework of definitions, the panel adopted an approach that evaluates how seriously the various behaviors compromise the integrity of the research process. The panel also considered other criteria, such as intent to deceive. The panel concluded that while intention is important, especially in the adjudication of allegations of misconduct in science, intention is often hard to establish and does not provide, by itself, an adequate basis for separating actions that seriously damage the integrity of the research process from questionable research practices or other misconduct.[16][17]

Misconduct in Science

Misconduct in science is defined as fabrication, falsification, or plagiarism, in proposing, performing, or reporting research. Misconduct in science does not include errors of judgment; errors in the recording, selection, or analysis of data; differences in opinions involving the interpretation of data; or misconduct unrelated to the research process.

Fabrication is making up data or results, falsification is changing data or results, and plagiarism is using the ideas or words of another person without giving appropriate credit.

By proposing this precise definition of misconduct in science, the panel is in unanimous agreement that the core of the definition of misconduct in science should consist of fabrication, falsification, and plagiarism. The panel unanimously rejects ambiguous language such as the category "other serious deviations from accepted research practices" currently included in regulatory definitions adopted by the Public Health Service and the National Science Foundation (DHHS, 1989a; NSF, 1991b). Although government officials have often relied on scientific panels to define "other serious deviations," the vagueness of this category has led to confusion about which actions constitute misconduct in science. In particular, the panel wishes to discourage the possibility that a misconduct complaint could be lodged against scientists based solely on their use of novel or unorthodox research methods. The use of ambiguous terms in regulatory definitions invites exactly such an overexpansive interpretation.

In rejecting the "other serious deviations" category, the panel considered whether a different measure of flexibility should be included in its proposed definition of misconduct in science, so as to allow the imposition of sanctions for conduct similar in character to fabrication, falsification, and plagiarism. Some panel members believe that the definition should also encompass other actions that directly damage the integrity of the research process and that are undertaken with the intent to deceive. For example, misuse of the peer-review system to penalize competitors, deceptive selection of data or statistical analysis, or encouragement of trainees to practice misconduct in science might not always constitute a form of fabrication, falsification, or plagiarism. Yet such actions could, in some circumstances, damage the integrity of the research process sufficiently to constitute misconduct in science.

All members of the panel support the basic definition of misconduct in science proposed above, but the panel did not reach final consensus on whether additional flexibility was needed to address as misconduct in science other practices of an egregious character similar to fabrication, falsification, and plagiarism. These issues deserve further consideration by the scientific research community to determine whether the panel's definition of misconduct in science is flexible enough to include all or most actions that directly damage the integrity of the research process and that were undertaken with the intent to deceive.

Questionable Research Practices

Questionable research practices are actions that violate traditional values of the research enterprise and that may be detrimental to the research process. However, there is at present neither broad agreement as to the seriousness of these actions nor any consensus on standards for behavior in such matters. Questionable research practices do not directly damage the integrity of the research process and thus do not meet the panel's criteria for inclusion in the definition of misconduct in science. However, they deserve attention because they can erode confidence in the integrity of the research process, violate traditions associated with science, affect scientific conclusions, waste time and resources, and weaken the education of new scientists.

Questionable research practices include activities such as the following:

- Failing to retain significant research data for a reasonable period;
- Maintaining inadequate research records, especially for results that are published or are relied on by others;
- Conferring or requesting authorship on the basis of a specialized service or contribution that is not significantly related to the research reported in the paper;[18]
- Refusing to give peers reasonable access to unique research materials or data that support published papers;
- Using inappropriate statistical or other methods of measurement to enhance the significance of research findings;[19]
- Inadequately supervising research subordinates or exploiting them; and
- Misrepresenting speculations as fact or releasing preliminary research results, especially in the public media, without providing sufficient data to allow peers to judge the validity of the results or to reproduce the experiments.

The panel wishes to make a clear demarcation between misconduct in science and questionable research practices—the two categories are not equivalent, and they require different types of responses by the research community and research institutions. However, the relationship between these two categories is not well understood. It may be difficult to tell, initially, whether alleged misconduct constitutes misconduct in science or a questionable research practice. In some cases, for example, scientists accused of plagiarism have testified about an absence of appropriate training methods for properly citing the work of others. The selective use of research data is another area where the boundary between fabrication and creative insight may not be obvious.

The panel emphasizes that scientists, individually and collectively, need to take questionable research practices seriously because when tolerated, such practices can encourage an environment that fosters misconduct in science. But questionable practices are not equivalent to misconduct in science, and they are not appropriate subjects for investigations directed to misconduct.

Other Misconduct

Certain forms of unacceptable behavior are clearly not unique to the conduct of science, although they may occur in a laboratory or research environment. Such behaviors, which are subject to generally applicable legal and social penalties, include actions such as sexual and other forms of harassment of individuals; misuse of funds; gross negligence by persons in their professional activities; vandalism, including tampering with research experiments or instrumentation;[20] and violations of government research regulations, such as those dealing with radioactive materials, recombinant DNA research, and the use of human or animal subjects. Industry-university relationships, and the resultant possibility of conflicts of interest, also raise issues that require special attention.

In these cases, recognized legal and institutional procedures should be in place to address complaints and to discourage behavior involving forms of misconduct that are not unique to the research process. Allegations of harassment, for example, should be handled by officials designated to implement personnel or equal opportunity regulations. Allegations of misuse of research funds should be addressed by those responsible for the financial integrity of the research institutions involved. The panel concluded that such behaviors require serious attention but lie outside the scope of the charge for this study.

On some occasions, however, certain forms of "other misconduct" are directly associated with misconduct in science. Among these are cover-ups of misconduct in science, reprisals against whistle-blowers, malicious allegations of misconduct in science, and violations of due process protections in handling complaints of misconduct in science. These forms of other misconduct may require action and special administrative procedures (see Chapter 5 for further discussion).

Understanding Causes and Evaluating Cures

The causes of misconduct in science are undoubtedly diverse and complex. Individual scientists, institutional officials, and scholars in the social studies of science over the past decade have suggested that various factors lead to or encourage misconduct in science, but the influence of any individual factor or combination of suggested factors has not been examined systematically.

Two alternate, possibly complementary, hypotheses have been advanced for considering the causes of misconduct in science and formulating methods for prevention and treatment. Many observers have explained the problem of misconduct in science as one that results primarily from character or personality flaws, from environmental stimuli in the research system, or from some interaction of both:[21]

1. *Misconduct in science is the result of individual pathology.* Misconduct in science is commonly viewed as the action of a psychologically disturbed individual. An analysis by Bechtel and Pearson (1985) of 12 cases of deviant behavior reported in the 1970s and early 1980s supported the hypothesis that scientists who engage in deviant behavior are commonly individuals who operate alone and who conceal their misconduct.[22]

2. *Factors in the modern research environment contribute to misconduct in science.* But although the "bad person" approach to explaining deviant behavior in science has had strong support within the scientific community, Bechtel and Pearson and others have questioned whether this hypothesis alone adequately explains the phenomenon of misconduct in science.

A broad range of factors in the research environment have been suggested as possible causes of misconduct in science. Such factors include (a) funding and career pressures of the contemporary research environment (such as the pressure to publish; NSB, 1988); (b) inadequate institutional oversight; (c) inappropriate forms of collaborative arrangements between academic scientists and commercial firms; (d) inadequate training in the methods and traditions of science;[23] (e) the increasing scale and complexity of the research environment, leading to the erosion of peer review, mentorship, and educational processes in science; and (f) the possibility that misconduct in science is an expression of a broader social pattern of deviation from traditional norms. In addition, it has been noted that some areas of research, such as biological and clinical research, do not yet proceed from explicit scientific laws and also make extensive use of empirical observations not related to theory. Moreover, the characteristics of certain research materials in these fields inhibit the replication of research findings as a vehicle for self-correction.

The panel has reviewed various suggestions about possible causes of misconduct in science but makes no judgment about the significance of any one factor. The panel believes that speculations about individual pathology or about environmental factors as the primary causes have not been verified; misconduct in science is probably the result of a complicated interaction of psychological and environmental factors. Moreover, although one or more such factors may contribute to specific cases of misconduct in science, the panel has not discerned a broad trend that would highlight any single factor as a clear generic cause.

Regardless of the causes of deviant behavior, the panel is concerned that some "cures" for misconduct in science would damage the research process itself. The uncertainty of evidence about external factors as causes means that recommending policy solutions for treating and preventing the problem of misconduct in science is problematic. As a result, efforts to foster integrity in the research process and to reduce the occurrence of misconduct in science should be evaluated systematically to identify steps that prove to be effective. A range of possible steps is discussed in the following chapters.

Starting from Logical Assumptions

The integrity of the research process has sometimes been called into question by sensationalized reports about specific cases of misconduct in science.[24] But because misconduct in science seems infrequent, many scientists have suggested that it does not present a serious problem. According to this view, when misconduct occurs in an important field of research, incorrect information will be corrected or eventually replaced by correct results through the work of others.

The panel agrees that confirmed cases of misconduct in science are rare. Nevertheless, the panel believes that every case of misconduct in science is serious and requires action for the following reasons:

1. *Misconduct is wrong.* One can object to misconduct in science simply on ethical grounds, since it often involves actions that betray personal and public trust and the search for truth. Misconduct in

science, if not properly addressed, can undermine the reasons for doing and supporting science itself.

2. *Misconduct in science wastes time and resources.* Misconduct can mislead scientists and waste the efforts of those who try to build on reported results. It requires substantial effort to correct false claims. Plagiarism can discourage scientists who see their contributions stolen or misrepresented by others and can damage honest reputations and the intellectual audit trail that affects the history of science.

3. *Misconduct can lead to injuries and harmful consequences.* Significant harm can result if false claims influence public health or technical or political decisions. Although mechanisms of self-correction may expose false claims, they are not designed to detect or deter misconduct in science. False information relating to medical procedures, for example, may lead to mistreatment of patients. Falsehoods should be publicly corrected, as soon as possible, to prevent such damage. We should not wait for the slow corrective action of further research. Similar comments apply in other areas of science in which false reports may have adverse practical consequences.

The time interval between the release and application of initial research reports in medical treatment, commercial products, services, and public policy decisions is diminishing. Resources for replicative research may not be available in some areas of research. Thus correction of research results, through replicative or related research efforts, is not a panacea; neither is it always timely.

4. *Misconduct by scientists, and weak institutional responses to these incidents, can lead to counter-productive regulation and control.* The image of scientists cheating in their laboratories is deeply disturbing to scientists themselves and to members of the public who have generally held scientists in high esteem. Even a few well-publicized cases of misconduct in science, particularly when such cases involve prominent individuals at respected institutions, have stimulated legal and administrative demands for accountability that divert funds and attention from scholarly purposes, interfere with the traditional autonomy granted to science, and malign the status of reputable scientists and their institutions.

5. *Misconduct in science can undermine public support of science.* Misconduct is one part of a larger public examination of scientific and educational institutions. Public confidence in the methods by which scientists maintain the integrity of the research process can be eroded when misconduct occurs in a social environment that is already disturbed by, for example, reports of misuse of the indirect costs associated with research funds, and other behaviors that violate public trust.

On the basis of these assumptions, the panel concluded that actions designed both to foster the integrity of the research process and to respond to misconduct in science are both timely and warranted.

NOTES

1. The values that characterize science are discussed in National Academy of Sciences (1989).

2. See, for example, further discussion on the ethos of science as described in Chapter 12 in Holton (1988). See also Sigma Xi (1986).

3. For a review of the impact of the contemporary research environment on the ethos of science, see Hoshiko (1991).

4. Government funding for U.S. basic research increased in current dollars from $5.4 billion in FY 1982 to an estimated $12.5 billion in FY 1991. See p. 53 in American Association for the Advancement of Science (1991a).

Academic research investigators are also increasingly supported by nonfederal funds provided by a diverse mix of industrial sponsors, state, and local funds, foundations, and intramural support. For example, the industrial share of academic R&D funding grew from 3.9 percent in 1980 to an estimated 6.6 percent in 1989. Some specialized academic research centers now receive over 20 percent of their funding from industry. See p. 106 in National Science Board (1989).

5. These factors include competitive pressures to publish, increasing competition for funds, secrecy in research performance, and inadequate interaction of young researchers with their peers and mentors. See Institute of Medicine (1989a).

6. See, for example, the following statement of Rep. John Dingell: "We are directing our efforts to seeing to it that NIH is able to function efficiently, well, honorably and competently in the public interest. We expect them to do that with full attention to their responsibilities to the taxpayers, as well as their duties towards the achievement of good science" (U.S. Congress, 1990c, p. 4).

7. As noted in *On Being a Scientist* (NAS, 1989), Alexander Kohn (1986) presents several case studies of fraud and self-deception from the history of science and medicine. A more popularly written and controversial history of misconduct in science is presented in Broad and Wade (1982).

Individual case histories have been reported in various journals and in newspaper accounts. See, for example, a summary of the controversy surrounding William Summerlin in McBride (1974) and an account of the Long, Soman, Alsabti, Straus, and Burt cases in Broad (1981).

8. See, for example, the cases described by Mazur (1989). See also the discussions in congressional oversight hearings (including U.S. Congress, 1981a; 1988a,b,c).

9. The term "allegation" here refers to complaints of misconduct in science that have resulted in a government case file. An analysis of these allegations is provided in Chapter 4. As of December 1991, about half of these allegations had been resolved.

10. Health Research Extension Act of 1985, P.L. 100-504, 99 Stat. 820 (1985).

11. See, for example, the reports resulting from three workshops sponsored by the National Conference of Lawyers and Scientists, American Association for the Advancement of Science and the American Bar Association (AAAS-ABA, 1989).

12. Some good examples of studies of scientific practice and the social organization of science include Traweek (1988), Hull (1988), Latour (1987), Latour and Woolgar (1979), Hackett and Chubin (1990), and Hackett (1990).

13. It is the panel's hope that the base of knowledge will be augmented by addi-

tional data derived from systematic evaluation of experiences in fostering responsible research practices. See also in Volume II of this report the background paper on this topic prepared for the panel by Nicholas Steneck.

14. A discussion of the dimensions of integrity in science is included in chapters 1 and 12 in Holton (1988).

15. Discussions focused initially on "scientific fraud" but encountered difficulties with the legal definition of the word "fraud." Government regulations and institutional policies have adopted terms such as "research misconduct," "scientific misconduct," and "misconduct in science," but these terms are subject to a variety of interpretations.

For early discussions about the relationship between fraud and misconduct in science, see Andersen (1988). See also the discussion on "fraud" and "misconduct" on p. 32447 in Department of Health and Human Services (1989a).

Some scientists object to the terms "scientific fraud" or "misconduct in science" because the fabrication and falsification of research results are deceptive acts that are not in themselves science. However, the social, political, and legal framework in which scientists must operate requires that we admit to the possibility of deliberate falsehoods that may masquerade as science.

16. Some institutional policies make *intention* or *deception* an explicit part of their definition of misconduct in science, whereas other policies assume, implicitly, that intention is part of the common understanding of actions, such as falsification, fabrication, and plagiarism, that constitute misconduct in science. See, for example, the definitions in the policies for addressing allegations of misconduct in science included in Volume II of this report.

17. Another approach considered by the panel in defining behaviors that violate the integrity of the research process was to deal only with misconduct in science and questionable research practices and to omit "other misconduct" as a category for a framework of definitions. Although the panel chose to focus on behaviors that directly compromise the integrity of the research process, it also wanted to recognize the public dimensions of discussions about misconduct in science. Thus the panel concluded that issues such as conflict of interest, mismanagement of funds, and the harassment of colleagues on the basis of race or gender must necessarily be recognized in a framework of definitions intended to categorize behavior that adversely affects the conduct of scientific research. These forms of "other misconduct" deserve serious and sustained analysis on their own merits, but such an examination was beyond the resources and scope of this particular study.

18. It is possible that some extreme cases of noncontributing authorship may be regarded as misconduct in science because they constitute a form of falsification. These would include only cases in which an individual who has made *no* identifiable contribution to a research paper is named, or seeks to be named, as a co-author.

19. See Bailar (1986).

20. The fourth report of the NSF inspector general (NSF, 1991a) describes a misconduct case involving tampering with other researchers' experiments. This type of case would not constitute misconduct in science under the panel's definition. An allegation of this type of incident should be addressed under regulations governing vandalism or destruction of property.

21. As noted in Bechtel and Pearson (1985), several leading figures in the scientific community have advocated the "disturbed individual" theory.

For discussions of the impact of reward systems and social controls on deviant behavior in science, see the analysis by Zuckerman (1977). For a historical perspective, see Gaston (1978).

22. The authors concluded that the deviant behavior in these cases, usually faking scientific experiments and data, was displayed by single individuals who acted alone. They observed that many of these individuals held positions of high social status and respectability within their professions and that the scientists involved also made elaborate efforts to conceal their illegitimate behavior.

23. It has been suggested that research physicians whose sole degree is an M.D. have not been adequately exposed to the scientific methods and skills that are the foundation of a Ph.D. program.

24. See Broad and Wade (1982).

2

Scientific Principles and Research Practices

Until the past decade, scientists, research institutions, and government agencies relied solely on a system of self-regulation based on shared ethical principles and generally accepted research practices to ensure integrity in the research process. Among the very basic principles that guide scientists, as well as many other scholars, are those expressed as respect for the integrity of knowledge, collegiality, honesty, objectivity, and openness. These principles are at work in the fundamental elements of the scientific method, such as formulating a hypothesis, designing an experiment to test the hypothesis, and collecting and interpreting data. In addition, more particular principles characteristic of specific scientific disciplines influence the methods of observation; the acquisition, storage, management, and sharing of data; the communication of scientific knowledge and information; and the training of younger scientists.[1] How these principles are applied varies considerably among the several scientific disciplines, different research organizations, and individual investigators.

The basic and particular principles that guide scientific research practices exist primarily in an unwritten code of ethics. Although some have proposed that these principles should be written down and formalized,[2] the principles and traditions of science are, for the most part, conveyed to successive generations of scientists through example, discussion, and informal education. As was pointed out in an early Academy report on responsible conduct of research in the health sciences, "a variety of informal and formal practices and procedures currently exist in the academic research environment to assure and maintain the high quality of research conduct" (IOM, 1989a, p. 18).

Physicist Richard Feynman invoked the informal approach to communicating the basic principles of science in his 1974 commencement address at the California Institute of Technology (Feynman, 1985):

> [There is an] idea that we all hope you have learned in studying science in school—we never explicitly say what this is, but just hope that you catch on by all the examples of scientific investigation. . . . It's a kind of scientific integrity, a principle of scientific thought that corresponds to a kind of utter honesty—a kind of leaning over backwards. For example, if you're doing an experiment, you should report everything that you think might make it invalid—not only what you think is right about it; other causes that could possibly explain your results; and things you thought of that you've eliminated by some other experiment, and how they worked—to make sure the other fellow can tell they have been eliminated.
>
> Details that could throw doubt on your interpretation must be given, if you know them. You must do the best you can—if you know anything at all wrong, or possibly wrong—to explain it. If you make a theory, for example, and advertise it, or put it out, then you must also put down all the facts that disagree with it, as well as those that agree with it. In summary, the idea is to try to give *all* the information to help others to judge the value of your contribution, not just the information that leads to judgment in one particular direction or another. (pp. 311-312)

Many scholars have noted the implicit nature and informal character of the processes that often guide scientific research practices and inference.[3] Research in well-established fields of scientific knowledge, guided by commonly accepted theoretical paradigms and experimental methods, involves few disagreements about what is recognized as sound scientific evidence. Even in a revolutionary scientific field like molecular biology, students and trainees have learned the basic principles governing judgments made in such standardized procedures as cloning a new gene and determining its sequence.

In evaluating practices that guide research endeavors, it is important to consider the individual character of scientific fields. Research fields that yield highly replicable results, such as ordinary organic chemical structures, are quite different from fields such as cellular immunology, which are in a much earlier stage of development and accumulate much erroneous or uninterpretable material before the pieces fit together coherently. When a research field is too new or

THE NATURE OF SCIENCE

In broadest terms, scientists seek a systematic organization of knowledge about the universe and its parts. This knowledge is based on explanatory principles whose verifiable consequences can be tested by independent observers. Science encompasses a large body of evidence collected by repeated observations and experiments. Although its goal is to approach true explanations as closely as possible, its investigators claim no final or permanent explanatory truths. Science changes. It evolves. Verifiable facts always take precedence. . . .

Scientists operate within a system designed for continuous testing, where corrections and new findings are announced in refereed scientific publications. The task of systematizing and extending the understanding of the universe is advanced by eliminating disproved ideas and by formulating new tests of others until one emerges as the most probable explanation for any given observed phenomenon. This is called the scientific method.

An idea that has not yet been sufficiently tested is called a hypothesis. Different hypotheses are sometimes advanced to explain the same factual evidence. Rigor in the testing of hypotheses is the heart of science. If no verifiable tests can be formulated, the idea is called an *ad hoc* hypothesis—one that is not fruitful; such hypotheses fail to stimulate research and are unlikely to advance scientific knowledge.

A fruitful hypothesis may develop into a theory after substantial observational or experimental support has accumulated. When a hypothesis has survived repeated opportunities for disproof and when competing hypotheses have been eliminated as a result of failure to produce the predicted consequences, that hypothesis may become the accepted theory explaining the original facts.

Scientific theories are also predictive. They allow us to anticipate yet unknown phenomena and thus to focus research on more narrowly defined areas. If the results of testing agree with predictions from a theory, the theory is provisionally corroborated. If not, it is proved false and must be either abandoned or modified to account for the inconsistency.

Scientific theories, therefore, are accepted only provisionally. It is always possible that a theory that has withstood previous testing may eventually be disproved. But as theories survive more tests, they are regarded with higher levels of confidence. . . .

In science, then, facts are determined by observation or measurement of natural or experimental phenomena. A hypothesis is a proposed explanation of those facts. A theory is a hypothesis that has gained wide acceptance because it has survived rigorous investigation of its predictions. . . .

continued

. . . science accommodates, indeed welcomes, new discoveries: its theories change and its activities broaden as new facts come to light or new potentials are recognized. Examples of events changing scientific thought are legion. . . . Truly scientific understanding cannot be attained or even pursued effectively when explanations not derived from or tested by the scientific method are accepted.

SOURCE: National Academy of Sciences and National Research Council (1984), pp. 8-11.

too fragmented to support consensual paradigms or established methods, different scientific practices can emerge.

A well-established discipline can also experience profound changes during periods of new conceptual insights. In these moments, when scientists must cope with shifting concepts, the matter of what counts as scientific evidence can be subject to dispute. Historian Jan Sapp has described the complex interplay between theory and observation that characterizes the operation of scientific judgment in the selection of research data during revolutionary periods of paradigmatic shift (Sapp, 1990, p. 113):

What "liberties" scientists are allowed in selecting positive data and omitting conflicting or "messy" data from their reports is not defined by any timeless method. It is a matter of negotiation. It is learned, acquired socially; scientists make judgments about what fellow scientists might expect in order to be convincing. What counts as good evidence may be more or less well-defined after a new discipline or specialty is formed; however, at revolutionary stages in science, when new theories and techniques are being put forward, when standards have yet to be negotiated, scientists are less certain as to what others may require of them to be deemed competent and convincing.

Explicit statements of the values and traditions that guide research practice have evolved through the disciplines and have been given in textbooks on scientific methodologies.[4] In the past few decades, many scientific and engineering societies representing individual disciplines have also adopted codes of ethics (see Volume II of this report for examples),[5] and more recently, a few research institutions have developed guidelines for the conduct of research (see Chapter 6).

But the responsibilities of the research community and research institutions in assuring individual compliance with scientific principles, traditions, and codes of ethics are not well defined. In recent years, the absence of formal statements by research institutions of the principles that should guide research conducted by their members has prompted criticism that scientists and their institutions lack a clearly identifiable means to ensure the integrity of the research process.

FACTORS AFFECTING THE DEVELOPMENT OF RESEARCH PRACTICES

In all of science, but with unequal emphasis in the several disciplines, inquiry proceeds based on observation and experimentation, the exercising of informed judgment, and the development of theory. Research practices are influenced by a variety of factors, including:

1. The general norms of science;
2. The nature of particular scientific disciplines and the traditions of organizing a specific body of scientific knowledge;
3. The example of individual scientists, particularly those who hold positions of authority or respect based on scientific achievements;
4. The policies and procedures of research institutions and funding agencies; and
5. Socially determined expectations.

The first three factors have been important in the evolution of modern science. The latter two have acquired more importance in recent times.

Norms of Science

As members of a professional group, scientists share a set of common values, aspirations, training, and work experiences.[6] Scientists are distinguished from other groups by their beliefs about the kinds of relationships that should exist among them, about the obligations incurred by members of their profession, and about their role in society. A set of general norms are imbedded in the methods and the disciplines of science that guide individual scientists in the organization and performance of their research efforts and that also provide a basis for nonscientists to understand and evaluate the performance of scientists.

But there is uncertainty about the extent to which individual scientists adhere to such norms. Most social scientists conclude that all behavior is influenced to some degree by norms that reflect socially or morally supported patterns of preference when alternative courses of action are possible. However, perfect conformity with any relevant set of norms is always lacking for a variety of reasons: the existence of competing norms, constraints and obstacles in organizational or group settings, and personality factors. The strength of these influences, and the circumstances that may affect them, are not well understood.

In a classic statement of the importance of scientific norms, Robert Merton specified four norms as essential for the effective functioning of science: communism (by which Merton meant the communal sharing of ideas and findings), universalism, disinterestedness, and organized skepticism (Merton, 1973). Neither Merton nor other sociologists of science have provided solid empirical evidence for the degree of influence of these norms in a representative sample of scientists. In opposition to Merton, a British sociologist of science, Michael Mulkay, has argued that these norms are "ideological" covers for self-interested behavior that reflects status and politics (Mulkay, 1975). And the British physicist and sociologist of science John Ziman, in an article synthesizing critiques of Merton's formulation, has specified a set of structural factors in the bureaucratic and corporate research environment that impede the realization of that particular set of norms: the proprietary nature of research, the local importance and funding of research, the authoritarian role of the research manager, commissioned research, and the required expertise in understanding how to use modern instruments (Ziman, 1990).

It is clear that the specific influence of norms on the development of scientific research practices is simply not known and that further study of key determinants is required, both theoretically and empirically. Commonsense views, ideologies, and anecdotes will not support a conclusive appraisal.

Individual Scientific Disciplines

Science comprises individual disciplines that reflect historical developments and the organization of natural and social phenomena for study. Social scientists may have methods for recording research data that differ from the methods of biologists, and scientists who depend on complex instrumentation may have authorship practices different from those of scientists who work in small groups or carry out field studies. Even within a discipline, experimentalists engage in research practices that differ from the procedures followed by theorists.

Disciplines are the "building blocks of science," and they "designate the theories, problems, procedures, and solutions that are prescribed, proscribed, permitted, and preferred" (Zuckerman, 1988a,

p. 520). The disciplines have traditionally provided the vital connections between scientific knowledge and its social organization. Scientific societies and scientific journals, some of which have tens of thousands of members and readers, and the peer review processes used by journals and research sponsors are visible forms of the social organization of the disciplines.

The power of the disciplines to shape research practices and standards is derived from their ability to provide a common frame of reference in evaluating the significance of new discoveries and theories in science. It is the members of a discipline, for example, who determine what is "good biology" or "good physics" by examining the implications of new research results. The disciplines' abilities to influence research standards are affected by the subjective quality of peer review and the extent to which factors other than disciplinary quality may affect judgments about scientific achievements. Disciplinary departments rely primarily on informal social and professional controls to promote responsible behavior and to penalize deviant behavior. These controls, such as social ostracism, the denial of letters of support for future employment, and the withholding of research resources, can deter and penalize unprofessional behavior within research institutions.[7]

Many scientific societies representing individual disciplines have adopted explicit standards in the form of codes of ethics or guidelines governing, for example, the editorial practices of their journals and other publications.[8] Many societies have also established procedures for enforcing their standards. In the past decade, the societies' codes of ethics—which historically have been exhortations to uphold high standards of professional behavior—have incorporated specific guidelines relevant to authorship practices, data management, training and mentoring, conflict of interest, reporting research findings, treatment of confidential or proprietary information, and addressing error or misconduct.

The Role of Individual Scientists and Research Teams

The methods by which individual scientists and students are socialized in the principles and traditions of science are poorly understood. The principles of science and the practices of the disciplines are transmitted by scientists in classroom settings and, perhaps more importantly, in research groups and teams. The social setting of the research group is a strong and valuable characteristic of American science and education. The dynamics of research groups can foster— or inhibit—innovation, creativity, education, and collaboration.

One author of a historical study of research groups in the chemical and biochemical sciences has observed that the laboratory director or group leader is the primary determinant of a group's practices (Fruton, 1990). Individuals in positions of authority are visible and are also influential in determining funding and other support for the career paths of their associates and students. Research directors and department chairs, by virtue of personal example, thus can reinforce, or weaken, the power of disciplinary standards and scientific norms to affect research practices.

To the extent that the behavior of senior scientists conforms with general expectations for appropriate scientific and disciplinary practice, the research system is coherent and mutually reinforcing. When the behavior of research directors or department chairs diverges from expectations for good practice, however, the expected norms of science become ambiguous, and their effects are thus weakened. Thus personal example and the perceived behavior of role models and leaders in the research community can be powerful stimuli in shaping the research practices of colleagues, associates, and students.

The role of individuals in influencing research practices can vary by research field, institution, or time. The standards and expectations for behavior exemplified by scientists who are highly regarded for their technical competence or creative insight may have greater influence than the standards of others. Individual and group behaviors may also be more influential in times of uncertainty and change in science, especially when new scientific theories, paradigms, or institutional relationships are being established.

Institutional Policies

Universities, independent institutes, and government and industrial research organizations create the environment in which research is done. As the recipients of federal funds and the institutional sponsors of research activities, administrative officers must comply with regulatory and legal requirements that accompany public support. They are required, for example, "to foster a research environment that discourages misconduct in all research and that deals forthrightly with possible misconduct" (DHHS, 1989a, p. 32451).

Academic institutions traditionally have relied on their faculty to ensure that appropriate scientific and disciplinary standards are maintained. A few universities and other research institutions have also adopted policies or guidelines to clarify the principles that their members are expected to observe in the conduct of scientific research.[9] In addition, as a result of several highly publicized incidents of misconduct in science and the subsequent enactment of governmental regulations, most major research institutions have now adopted policies and procedures for handling allegations of misconduct in science.

Institutional policies governing research practices can have a powerful effect on research practices if they are commensurate with the norms that apply to a wide spectrum of research investigators. In particular, the process of adopting and implementing strong institutional policies can sensitize the members of those institutions to the potential for ethical problems in their work. Institutional policies can establish explicit standards that institutional officers then have the power to enforce with sanctions and penalties.

Institutional policies are limited, however, in their ability to specify the details of every problematic situation, and they can weaken or displace individual professional judgment in such situations. Currently, academic institutions have very few formal policies and programs in specific areas such as authorship, communication and publication, and training and supervision.

Government Regulations and Policies

Government agencies have developed specific rules and procedures that directly affect research practices in areas such as laboratory safety, the treatment of human and animal research subjects, and the use of toxic or potentially hazardous substances in research.

But policies and procedures adopted by some government research agencies to address misconduct in science (see Chapter 5) represent a significant new regulatory development in the relationships between research institutions and government sponsors. The standards and criteria used to monitor institutional compliance with an increasing number of government regulations and policies affecting research practices have been a source of significant disagreement and tension within the research community.

In recent years, some government research agencies have also adopted policies and procedures for the treatment of research data and materials in their extramural research programs. For example, the National Science Foundation (NSF) has implemented a data-sharing policy through program management actions, including proposal review and award negotiations and conditions. The NSF policy acknowledges that grantee institutions will "keep principal rights to intellectual property conceived under NSF sponsorship" to encourage appropriate commercialization of the results of research (NSF, 1989b, p. 1). However, the NSF policy emphasizes "that retention of such rights does not reduce the responsibility of researchers and institutions to make results and supporting materials openly accessible" (p. 1).

In seeking to foster data sharing under federal grant awards, the government relies extensively on the scientific traditions of openness and sharing. Research agency officials have observed candidly that if the vast majority of scientists were not so committed to openness and dissemination, government policy might require more aggressive action. But the principles that have traditionally characterized scientific inquiry can be difficult to maintain. For example, NSF staff have commented, "Unless we can arrange real returns or incentives for the original investigator, either in financial support or in professional recognition, another researcher's request for sharing is likely to present itself as 'hassle'—an unwelcome nuisance and diversion. Therefore, we should hardly be surprised if researchers display some reluctance to share in practice, however much they may declare and genuinely feel devotion to the ideal of open scientific communication" (NSF, 1989a, p. 4).

Social Attitudes and Expectations

Research scientists are part of a larger human society that has recently experienced profound changes in attitudes about ethics, morality, and accountability in business, the professions, and government. These attitudes have included greater skepticism of the authority of experts and broader expectations about the need for visible mechanisms to assure proper research practices, especially in areas that affect the public welfare. Social attitudes are also having a more direct influence on research practices as science achieves a more prominent and public role in society. In particular, concern about waste, fraud, and abuse involving government funds has emerged as a factor that now directly influences the practices of the research community.

Varying historical and conceptual perspectives also can affect expectations about standards of research practice. For example, some journalists have criticized several prominent scientists, such as Mendel, Newton, and Millikan, because they "cut corners in order to make their theories prevail" (Broad and Wade, 1982, p. 35). The criticism suggests that all scientists at all times, in all phases of their work, should be bound by identical standards.

Yet historical studies of the social context in which scientific knowledge has been attained suggest that modern criticism of early scientific work often imposes contemporary standards of objectivity and empiricism that have in fact been developed in an evolutionary manner.[10] Holton has argued, for example, that in selecting data for publication, Millikan exercised creative insight in excluding unreliable data resulting from experimental error. But such practices, by today's standards, would not be acceptable without reporting the justification for omission of recorded data.

In the early stages of pioneering studies, particularly when fundamental hypotheses are subject to change, scientists must be free to use creative judgment in deciding which data are truly significant. In such moments, the standards of proof may be quite different from those that apply at stages when confirmation and consensus are sought from peers. Scientists must consistently guard against self-deception, however, particularly when theoretical prejudices tend to overwhelm the skepticism and objectivity basic to experimental practices.

In discussing "the theory-ladenness of observations," Sapp (1990) observed the fundamental paradox that can exist in determining the "appropriateness" of data selection in certain experiments done in the past: scientists often craft their experiments so that the scientific problems and research subjects conform closely with the theory that they expect to verify or refute. Thus, in some cases, their observations may come closer to theoretical expectations than what might be statistically proper.

This source of bias may be acceptable when it is influenced by scientific insight and judgment. But political, financial, or other sources of bias can corrupt the process of data selection. In situations where both kinds of influence exist, it is particularly important for scientists to be forthcoming about possible sources of bias in the interpretation of research results. The coupling of science to other social purposes in fostering economic growth and commercial technology requires renewed vigilance to maintain acceptable standards for disclosure and control of financial or competitive conflicts of interest and bias in the research environment. The failure to distinguish between appropriate and inappropriate sources of bias in research practices can lead to erosion of public trust in the autonomy of the research enterprise.

RESEARCH PRACTICES

In reviewing modern research practices for a range of disciplines, and analyzing factors that could affect the integrity of the research process, the panel focused on the following four areas:

1. Data handling—acquisition, management, and storage;
2. Communication and publication;

3. Correction of errors; and

4. Research training and mentorship.

Commonly understood practices operate in each area to promote responsible research conduct; nevertheless, some questionable research practices also occur. Some research institutions, scientific societies, and journals have established policies to discourage questionable practices, but there is not yet a consensus on how to treat violations of these policies.[11] Furthermore, there is concern that some questionable practices may be encouraged or stimulated by other institutional factors. For example, promotion or appointment policies that stress quantity rather than the quality of publications as a measure of productivity could contribute to questionable practices.

Data Handling

Acquisition and Management

Scientific experiments and measurements are transformed into research data. The term "research data" applies to many different forms of scientific information, including raw numbers and field notes, machine tapes and notebooks, edited and categorized observations, interpretations and analyses, derived reagents and vectors, and tables, charts, slides, and photographs.

Research data are the basis for reporting discoveries and experimental results. Scientists traditionally describe the methods used for an experiment, along with appropriate calibrations, instrument types, the number of repeated measurements, and particular conditions that may have led to the omission of some data in the reported version. Standard procedures, innovations for particular purposes, and judgments concerning the data are also reported. The general standard of practice is to provide information that is sufficiently complete so that another scientist can repeat or extend the experiment.

When a scientist communicates a set of results and a related piece of theory or interpretation in any form (at a meeting, in a journal article, or in a book), it is assumed that the research has been conducted as reported. It is a violation of the most fundamental aspect of the scientific research process to set forth measurements that have not, in fact, been performed (fabrication) or to ignore or change relevant data that contradict the reported findings (falsification).

On occasion what is actually proper research practice may be confused with misconduct in science. Thus, for example, applying scientific judgment to refine data and to remove spurious results places special responsibility on the researcher to avoid misrepresentation of findings. Responsible practice requires that scientists disclose the basis for omitting or modifying data in their analyses of research results, especially when such omissions or modifications could alter the interpretation or significance of their work.

In the last decade, the methods by which research scientists handle, store, and provide access to research data have received increased scrutiny, owing to conflicts over ownership, such as those described by Nelkin (1984); advances in the methods and technologies that are used to collect, retain, and share data; and the costs of data storage. More specific concerns have involved the profitability associated with the patenting of science-based results in some fields and the need to verify independently the accuracy of research results used in public or private decision making. In resolving competing claims, the interests of individual scientists and research institutions may not always coincide: researchers may be willing to exchange scientific data of possible economic significance without regard for financial or institutional implications, whereas their institutions may wish to establish intellectual property rights and obligations prior to any disclosure.

The general norms of science emphasize the principle of openness. Scientists are generally expected to exchange research data as well as unique research materials that are essential to the replication or extension of reported findings. The 1985 report *Sharing Research Data* concluded that the general principle of data sharing is widely accepted, especially in the behavioral and social sciences (NRC, 1985). The report catalogued the benefits of data sharing, including maintaining the integrity of the research process by providing independent opportunities for verification, refutation, or refinement of original results and data; promoting new research and the development and testing of new theories; and encouraging appropriate use of empirical data in policy formulation and evaluation. The same report examined obstacles to data sharing, which include the criticism or competition that might be stimulated by data sharing; technical barriers that may impede the exchange of computer-readable data; lack of documentation of data sets; and the considerable costs of documentation, duplication, and transfer of data.

The exchange of research data and reagents is ideally governed by principles of collegiality and reciprocity: scientists often distribute reagents with the hope that the recipient will reciprocate in the future, and some give materials out freely with no stipulations attached.[12] Scientists who repeatedly or flagrantly deviate from the tradition of sharing become known to their peers and may suffer subtle forms of professional isolation. Such cases may be well known to senior research investigators, but they are not well documented.

Some scientists may share materials as part of a collaborative agreement in exchange for co-authorship on resulting publications. Some donors stipulate that the shared materials are not to be used for applications already being pursued by the donor's laboratory. Other stipulations include that the material not be passed on to third parties without prior authorization, that the material not be used for proprietary research, or that the donor receive prepublication copies of research publications derived from the material. In some instances, so-called materials transfer agreements are executed to specify the responsibilities of donor and recipient. As more academic research is being supported under proprietary agreements, researchers and institutions are experiencing the effects of these arrangements on research practices.

Governmental support for research studies may raise fundamental questions of ownership and rights of control, particularly when data are subsequently used in proprietary efforts, public policy decisions, or litigation. Some federal research agencies have adopted policies for data sharing to mitigate conflicts over issues of ownership and access (NIH, 1987; NSF, 1989b).

Storage

Many research investigators store primary data in the laboratories in which the data were initially derived, generally as electronic records or data sheets in laboratory notebooks. For most academic laboratories, local customary practice governs the storage (or discarding) of research data. Formal rules or guidelines concerning their disposition are rare.

Many laboratories customarily store primary data for a set period (often 3 to 5 years) after they are initially collected. Data that support publications are usually retained for a longer period than are those tangential to reported results. Some research laboratories serve as the proprietor of data and data books that are under the stewardship of the principal investigator. Others maintain that it is the responsibility of the individuals who collected the data to retain proprietorship, even if they leave the laboratory.

Concerns about misconduct in science have raised questions about the roles of research investigators and of institutions in maintaining and providing access to primary data. In some cases of alleged misconduct, the inability or unwillingness of an investigator to provide primary data or witnesses to support published reports sometimes has constituted a presumption that the experiments were not conducted as reported.[13] Furthermore, there is disagreement about the responsibilities of investigators to provide access to raw data, particularly when the reported results have been challenged by others. Many scientists believe that access should be restricted to peers and colleagues, usually following publication of research results, to reduce external demands on the time of the investigator. Others have suggested that raw data supporting research reports should be accessible to any critic or competitor, at any time, especially if the research is conducted with public funds. This topic, in particular, could benefit from further research and systematic discussion to clarify the rights and responsibilities of research investigators, institutions, and sponsors.

Institutional policies have been developed to guide data storage practices in some fields, often stimulated by desires to support the patenting of scientific results and to provide documentation for resolving disputes over patent claims. Laboratories concerned with patents usually have very strict rules concerning data storage and note keeping, often requiring that notes be recorded in an indelible form and be countersigned by an authorized person each day. A few universities have also considered the creation of central storage repositories for all primary data collected by their research investigators. Some government research institutions and industrial research centers maintain such repositories to safeguard the record of research developments for scientific, historical, proprietary, and national security interests.

In the academic environment, however, centralized research records raise complex problems of ownership, control, and access. Centralized data storage is costly in terms of money and space, and it presents logistical problems of cataloguing and retrieving data. There have been suggestions that some types of scientific data should be incorporated into centralized computerized data banks, a portion of which could be subject to periodic auditing or certification.[14] But much investigator-initiated research is not suitable for random data audits because of the exploratory nature of basic or discovery research.[15]

Some scientific journals now require that full data for research papers be deposited in a centralized data bank before final publication. Policies and practices differ, but in some fields support is growing for compulsory deposit to enhance researchers' access to supporting data.

Issues Related to Advances in
Information Technology

Advances in electronic and other information technologies have raised new questions about the customs and practices that influence the storage, ownership, and exchange of electronic data and software. A number of special issues, not addressed by the panel, are associated with computer modeling, simulation, and other approaches that are becoming more prevalent in the research environment. Computer technology can enhance research collaboration; it can also create new impediments to data sharing resulting from increased costs, the need for specialized equipment, or liabilities or uncertainties about responsibilities for faulty data, software, or computer-generated models.

Advances in computer technology may assist in maintaining and preserving accurate records of research data. Such records could help resolve questions about the timing or accuracy of specific research findings, especially when a principal investigator is not available or is uncooperative in responding to such questions. In principle, properly managed information technologies, utilizing advances in nonerasable optical disk systems, might reinforce openness in scientific research and make primary data more transparent to collaborators and research managers. For example, the so-called WORM (write once, read many) systems provide a high-density digital storage medium that supplies an ineradicable audit trail and historical record for all entered information (Haas, 1991).

Advances in information technologies could thus provide an important benefit to research institutions that wish to emphasize greater access to and storage of primary research data. But the development of centralized information systems in the academic research environment raises difficult issues of ownership, control, and principle that reflect the decentralized character of university governance. Such systems are also a source of additional research expense, often borne by individual investigators. Moreover, if centralized systems are perceived by scientists as an inappropriate or ineffective form of management or oversight of individual research groups, they simply may not work in an academic environment.

Communication and Publication

Scientists communicate research results by a variety of formal and informal means. In earlier times, new findings and interpretations were communicated by letter, personal meeting, and publication. Today, computer networks and facsimile machines have supplemented letters and telephones in facilitating rapid exchange of results. Scientific meetings routinely include poster sessions and press conferences as well as formal presentations. Although research publications continue to document research findings, the appearance of electronic publications and other information technologies heralds change. In addition, incidents of plagiarism, the increasing number of authors per article in selected fields, and the methods by which publications are assessed in determining appointments and promotions have all increased concerns about the traditions and practices that have guided communication and publication.

Journal publication, traditionally an important means of sharing information and perspectives among scientists, is also a principal means of establishing a record of achievement in science. Evaluation of the accomplishments of individual scientists often involves not only the numbers of articles that have resulted from a selected research effort, but also the particular journals in which the articles have appeared. Journal submission dates are often important in establishing priority and intellectual property claims.

Authorship

Authorship of original research reports is an important indicator of accomplishment, priority, and prestige within the scientific community. Questions of authorship in science are intimately connected with issues of credit and responsibility. Authorship practices are guided by disciplinary traditions, customary practices within research groups, and professional and journal standards and policies.[16] There is general acceptance of the principle that each named author has made a significant intellectual contribution to the paper, even though there remains substantial disagreement over the types of contributions that are judged to be significant.

A general rule is that an author must have participated sufficiently in the work to take responsibility for its content and vouch for its validity. Some journals have adopted more specific guidelines, suggesting that credit for authorship be contingent on substantial participation in one or more of the following categories: (1) conception and design of the experiment, (2) execution of the experiment and collection and storage of the supporting data, (3) analysis and interpretation of the primary data, and (4) preparation and revision of the manuscript. The extent of participation in these four activities required for authorship varies across journals, disciplines, and research groups.[17]

"Honorary," "gift," or other forms of noncontributing authorship are problems with several dimensions.[18] Honorary authors reap an inflated list of publications incommensurate with their scientific contributions (Zen, 1988). Some scientists have requested or been given authorship as a form of recognition of their status or influence rather than their intellectual contribution. Some research leaders have a custom of including their own names in any paper issuing from their laboratory, although this practice is increasingly discouraged. Some students or junior staff encourage such "gift authorship" because they feel that the inclusion of prestigious names on their papers increases the chance of publication in well-known journals. In some cases, noncontributing authors have been listed without their consent, or even without their being told. In response to these practices, some journals now require all named authors to sign the letter that accompanies submission of the original article, to ensure that no author is named without consent.

"Specialized" authorship is another issue that has received increasing attention. In these cases, a co-author may claim responsibility for a specialized portion of the paper and may not even see or be able to defend the paper as a whole.[19] "Specialized" authorship may also result from demands that co-authorship be given as a condition of sharing a unique research reagent or selected data that do not constitute a major contribution—demands that many scientists believe are inappropriate. "Specialized" authorship may be appropriate in cross-disciplinary collaborations, in which each participant has made an important contribution that deserves recognition. However, the risks associated with the inabilities of co-authors to vouch for the integrity of an entire paper are great; scientists may unwittingly become associated with a discredited publication.

Another problem of lesser importance, except to the scientists involved, is the order of authors listed on a paper. The meaning of author order varies among and within disciplines. For example, in physics the ordering of authors is frequently alphabetical, whereas in the social sciences and other fields, the ordering reflects a descending order of contribution to the described research. Another practice, common in biology, is to list the senior author last.

Appropriate recognition for the contributions of junior investigators, postdoctoral fellows, and graduate students is sometimes a source of discontent and unease in the contemporary research environment. Junior researchers have raised concerns about treatment of their contributions when research papers are prepared and submitted, particularly if they are attempting to secure promotions or independent research funding or if they have left the original project. In some cases, well-meaning senior scientists may grant junior colleagues undeserved authorship or placement as a means of enhancing the junior colleague's reputation. In others, significant contributions may not receive appropriate recognition.

Authorship practices are further complicated by large-scale projects, especially those that involve specialized contributions. Mission teams for space probes, oceanographic expeditions, and projects in high-energy physics, for example, all involve large numbers of senior scientists who depend on the long-term functioning of complex equipment. Some questions about communication and publication that arise from large science projects such as the Superconducting Super Collider include: Who decides when an experiment is ready to be published? How is the spokesperson for the experiment determined? Who determines who can give talks on the experiment? How should credit for technical or hardware contributions be acknowledged?

Apart from plagiarism, problems of authorship and credit allocation usually do not involve misconduct in science. Although some forms of "gift authorship," in which a designated author made no identifiable contribution to a paper, may be viewed as instances of falsification, authorship disputes more commonly involve unresolved differences of judgment and style. Many research groups have found that the best method of resolving authorship questions is to agree on a designation of authors at the outset of the project. The negotiation and decision process provides initial recognition of each member's effort, and it may prevent misunderstandings that can arise during the course of the project when individuals may be in transition to new efforts or may become preoccupied with other matters.

Plagiarism. Plagiarism is using the ideas or words of another person without giving appropriate credit. Plagiarism includes the unacknowledged use of text and ideas from published work, as well as the misuse of privileged information obtained through confidential review of research proposals and manuscripts.

As described in *Honor in Science*, plagiarism can take many forms: at one extreme is the exact replication of another's writing without appropriate attribution (Sigma Xi, 1986). At the other is the more subtle "borrowing" of ideas, terms, or paraphrases, as described by Martin et al., "so that the result is a mosaic of other people's ideas and words, the writer's sole contribution being the cement to hold the pieces together."[20] The importance of recognition for one's intellectual abilities in science demands high standards of accuracy and diligence in ensuring appropriate recognition for the work of others.

The misuse of privileged information may be less clear-cut because it does not involve published work. But the general principles of the importance of giving credit to the accomplishments of others are the same. The use of ideas or information obtained from peer review is not acceptable because the reviewer is in a privileged position. Some organizations, such as the American Chemical Society, have adopted policies to address these concerns (ACS, 1986).

Additional Concerns. Other problems related to authorship include overspecialization, overemphasis on short-term projects, and the organization of research communication around the "least publishable unit." In a research system that rewards quantity at the expense of quality and favors speed over attention to detail (the effects of "publish or perish"), scientists who wait until their research data are complete before releasing them for publication may be at a disadvantage. Some institutions, such as Harvard Medical School, have responded to these problems by limiting the number of publications reviewed for promotion. Others have placed greater emphasis on major contributions as the basis for evaluating research productivity.

Editors

As gatekeepers of scientific journals, editors are expected to use good judgment and fairness in selecting papers for publication. Although editors cannot be held responsible for the errors or inaccuracies of papers that may appear in their journals, editors have obligations to consider criticism and evidence that might contradict the claims of an author and to facilitate publication of critical letters, errata, or retractions.[21] Some institutions, including the National Library of Medicine and professional societies that represent editors of scientific journals, are exploring the development of standards relevant to these obligations (Bailar et al., 1990).

Should questions be raised about the integrity of a published work, the editor may request an author's institution to address the matter. Editors often request written assurances that research reported conforms to all appropriate guidelines involving human or animal subjects, materials of human origin, or recombinant DNA.

In theory, editors set standards of authorship for their journals. In practice, scientists in the specialty do. Editors may specify the terms of acknowledgment of contributors who fall short of authorship status, and make decisions regarding appropriate forms of disclosure of sources of bias or other potential conflicts of interest related to published articles. For example, the *New England Journal of Medicine* has established a category of prohibited contributions from authors engaged in for-profit ventures: the journal will not allow such persons to prepare review articles or editorial commentaries for publication. Editors can clarify and insist on the confidentiality of review and take appropriate actions against reviewers who violate it. Journals also may require or encourage their authors to deposit reagents and sequence and crystallographic data into appropriate databases or storage facilities.[22]

Peer Review

Peer review is the process by which editors and journals seek to be advised by knowledgeable colleagues about the quality and suitability of a manuscript for publication in a journal. Peer review is also used by funding agencies to seek advice concerning the quality and promise of proposals for research support. The proliferation of research journals and the rewards associated with publication and with obtaining research grants have put substantial stress on the peer review system. Reviewers for journals or research agencies receive privileged information and must exert great care to avoid sharing such information with colleagues or allowing it to enter their own work prematurely.

Although the system of peer review is generally effective, it has been suggested that the quality of refereeing has declined, that self-interest has crept into the review process, and that some journal editors and reviewers exert inappropriate influence on the type of work they deem publishable.[23]

Correction of Errors

At some level, all scientific reports, even those that mark profound advances, contain errors of fact or interpretation. In part, such errors reflect uncertainties intrinsic to the research process itself—a hypothesis is formulated, an experimental test is devised, and based on the interpretation of the results, the hypothesis is refined, revised, or discarded. Each step in this cycle is subject to error. For any given report, "correctness" is limited by the following:

1. The precision and accuracy of the measurements. These in turn depend on available technology, the use of proper statistical and analytical methods, and the skills of the investigator.

2. Generality of the experimental system and approach. Studies must often be carried out using "model systems." In biology, for example, a given phenomenon is examined in only one or a few among millions of organismal species.

3. Experimental design—a product of the background and expertise of the investigator.

4. Interpretation and speculation regarding the significance of the findings—judgments that depend on expert knowledge, experience, and the insightfulness and boldness of the investigator.

Viewed in this context, errors are an integral aspect of progress in attaining scientific knowledge. They are consequences of the fact that scientists seek fundamental truths about natural processes of vast complexity. In the best experimental systems, it is common that relatively few variables have been identified and that even fewer can be controlled experimentally. Even when important variables are accounted for, the interpretation of the experimental results may be incorrect and may lead to an erroneous conclusion. Such conclusions are sometimes overturned by the original investigator or by others when new insights from another study prompt a reexamination of older reported data. In addition, however, erroneous information can also reach the scientific literature as a consequence of misconduct in science.

What becomes of these errors or incorrect interpretations? Much has been made of the concept that science is "self-correcting"—that errors, whether honest or products of misconduct, will be exposed in future experiments because scientific truth is founded on the principle that results must be verifiable and reproducible. This implies that errors will generally not long confound the direction of thinking or experimentation in actively pursued areas of research. Clearly, published experiments are not routinely replicated precisely by independent investigators. However, each experiment is based on conclusions from prior studies; repeated failure of the experiment eventually calls into question those conclusions and leads to reevaluation of the measurements, generality, design, and interpretation of the earlier work.

Thus publication of a scientific report provides an opportunity for the community at large to critique and build on the substance of the report, and serves as one stage at which errors and misinterpretations can be detected and corrected. Each new finding is considered by the community in light of what is already known about the system investigated, and disagreements with established measurements and interpretations must be justified. For example, a particular interpretation of an electrical measurement of a material may implicitly predict the results of an optical experiment. If the reported optical results are in disagreement with the electrical interpretation, then the latter is unlikely to be correct—even though the measurements themselves were carefully and correctly performed. It is also possible, however, that the contradictory results are themselves incorrect, and this possibility will also be evaluated by the scientists working in the field. It is by this process of examination and reexamination that science advances.

The research endeavor can therefore be viewed as a two-tiered process: first, hypotheses are formulated, tested, and modified; second, results and conclusions are reevaluated in the course of additional study. In fact, the two tiers are interrelated, and the goals and traditions of science mandate major responsibilities in both areas for individual investigators. Importantly, the principle of self-correction does not diminish the responsibilities of the investigator in either area. The investigator has a fundamental responsibility to ensure that the reported results can be replicated in his or her laboratory. The scientific community in general adheres strongly to this principle, but practical constraints exist as a result of the availability of specialized instrumentation, research materials, and expert personnel. Other forces, such as competition, commercial interest, funding trends and availability, or pressure to publish may also erode the role of replication as a mechanism for fostering integrity in the research process. The panel is unaware of any quantitative studies of this issue.

The process of reevaluating prior findings is closely related to the formulation and testing of hypotheses.[24] Indeed, within an individual laboratory, the formulation/testing phase and the reevaluation phase are ideally ongoing interactive processes. In that setting, the precise replication of a prior result commonly serves as a crucial control in attempts to extend the original findings. It is not unusual that experimental flaws or errors of interpretation are revealed as the scope of an investigation deepens and broadens.

If new findings or significant questions emerge in the course of a reevaluation that affect the claims of a published report, the investigator is obliged to make public a correction of the erroneous result or to indicate the nature of the questions. Occasionally, this takes the form of a formal published retraction, especially in situations in which a central claim is found to be fundamentally incorrect or irreproducible. More commonly, a somewhat different version of the original experiment, or a revised interpretation of the original result, is published as part of a subsequent report that extends in other ways the initial work. Some concerns have been raised that such "revisions" can sometimes be so subtle and obscure as to be unrecognizable. Such behavior is, at best, a questionable research practice. Clearly, each scientist has a responsibility to foster an environment that encourages and demands rigorous evaluation and reevaluation of every key finding.

Much greater complexity is encountered when an investigator in one research group is unable to confirm the published findings of another. In such situations, precise replication of the original result is commonly not attempted because of the lack of identical reagents, differences in experimental protocols, diverse experimental goals, or differences in personnel. Under these circumstances, attempts to obtain the published result may simply be dropped if the central claim of the original study is not the major focus of the new study. Alternatively, the inability to obtain the original finding may be documented in a paper by the second investigator as part of a challenge to the original claim. In any case, such questions about a published finding usually provoke the initial investigator to attempt to reconfirm the original result, or to pursue additional studies that support and extend the original findings.

In accordance with established principles of science, scientists have the responsibility to replicate and reconfirm their results as a normal part of the research process. The cycles of theoretical and methodological formulation, testing, and reevaluation, both within and between laboratories, produce an ongoing process of revision and refinement that corrects errors and strengthens the fabric of research.

Research Training and Mentorship

The panel defined a mentor as that person directly responsible for the professional development of a research trainee.[25] Professional development includes both technical training, such as instruction in the methods of scientific research (e.g., research design, instrument use, and selection of research questions and data), and socialization in basic research practices (e.g., authorship practices and sharing of research data).

Positive Aspects of Mentorship

The relationship of the mentor and research trainee is usually characterized by extraordinary mutual commitment and personal involvement. A mentor, as a research advisor, is generally expected to supervise the work of the trainee and ensure that the trainee's research is completed in a sound, honest, and timely manner. The ideal mentor challenges the trainee, spurs the trainee to higher scientific achievement, and helps socialize the trainee into the community of scientists by demonstrating and discussing methods and practices that are not well understood.

Research mentors thus have complex and diverse roles. Many individuals excel in providing guidance and instruction as well as personal support, and some mentors are resourceful in providing funds and securing professional opportunities for their trainees. The mentoring relationship may also combine elements of other relationships, such as parenting, coaching, and guildmastering. One mentor has written that his "research group is like an extended family or small tribe, dependent on one another, but led by the mentor, who acts as their consultant, critic, judge, advisor, and scientific father" (Cram, 1989, p. 1). Another mentor described as "orphaned graduate students" trainees who had lost their mentors to death, job changes, or in other ways (Sindermann, 1987). Many students come to respect and admire their mentors, who act as role models for their younger colleagues.

Difficulties Associated with Mentorship

However, the mentoring relationship does not always function properly or even satisfactorily. Almost no literature exists that evaluates which problems are idiosyncratic and which are systemic. However, it is clear that traditional practices in the area of mentorship and training are under stress. In some research fields, for example, concerns are being raised about how the increasing size and diverse composition of research groups affect the quality of the relationship between trainee and mentor. As the size of research laboratories expands, the quality of the training environment is at risk (CGS, 1990a).

Large laboratories may provide valuable instrumentation and access to unique research skills and resources as well as an opportunity to work in pioneering fields of science. But as only one contribution to the efforts of a large research team, a graduate student's work may become highly specialized, leading to a narrowing of experience and greater dependency on senior personnel; in a period when the availability of funding may limit research opportunities, laboratory heads may find it necessary to balance research decisions for the good of the team against the individual educational interests of each trainee. Moreover, the demands of obtaining sufficient resources to maintain a laboratory in the contemporary research environment often separate faculty from their trainees. When laboratory heads fail to participate in the everyday workings of the laboratory—even for the most beneficent of reasons, such as finding funds to support young investigators—their inattention may harm their trainees' education.

Although the size of a research group can influence the quality of mentorship, the more important issues are the level of supervision received by trainees, the degree of independence that is appropriate for the trainees' experience and interests, and the allocation of credit for achievements that are accomplished by groups composed of individuals with different status. Certain studies involving large groups of 40 to 100 or more are commonly carried out by collaborative or hierarchical arrangements under a single investigator. These factors may affect the ability of research mentors to transmit the methods and ethical principles according to which research should be conducted.

Problems also arise when faculty members are not directly rewarded for their graduate teaching or training skills. Although faculty may receive indirect rewards from the contributions of well-trained graduate students to their own research as well as the satisfaction of seeing their students excelling elsewhere, these rewards may not be sufficiently significant in tenure or promotion decisions. When institutional policies fail to recognize and reward the value of good teaching and mentorship, the pressures to maintain stable funding for research teams in a competitive environment can overwhelm the time allocated to teaching and mentorship by a single investigator.

The increasing duration of the training period in many research fields is another source of concern, particularly when it prolongs the dependent status of the junior investigator. The formal period of graduate and postdoctoral training varies considerably among fields of study. In 1988, the median time to the doctorate from the baccalaureate degree was 6.5 years (NRC, 1989). The disciplinary median varied: 5.5 years in chemistry; 5.9 years in engineering; 7.1 years in health sciences and in earth, atmospheric, and marine sciences; and 9.0 years in anthropology and sociology.[26]

Students, research associates, and faculty are currently raising various questions about the rights and obligations of trainees. Sexist behavior by some research directors and other senior scientists is a particular source of concern. Another significant concern is that research trainees may be subject to exploitation because of their subordinate status in the research laboratory, particularly when their income, access to research resources, and future recommendations are dependent on the goodwill of the mentor. Foreign students and postdoctoral fellows may be especially vulnerable, since their immigration status often depends on continuation of a research relationship with the selected mentor.

Inequalities between mentor and trainee can exacerbate ordinary conflicts such as the distribution of credit or blame for research error (NAS, 1989). When conflicts arise, the expectations and assumptions that govern authorship practices, ownership of intellectual property, and the giving of references and recommendations are exposed for professional—and even legal—scrutiny (Nelkin, 1984; Weil and Snapper, 1989).

Making Mentorship Better

Ideally, mentors and trainees should select each other with an eye toward scientific merit, intellectual and personal compatibility, and other relevant factors. But this situation operates only under conditions of freely available information and unconstrained choice— conditions that usually do not exist in academic research groups. The trainee may choose to work with a faculty member based solely on criteria of patronage, perceived influence, or ability to provide financial support.

Good mentors may be well known and highly regarded within their research communities and institutions. Unfortunately, individuals who exploit the mentorship relationship may be less visible. Poor mentorship practices may be self-correcting over time, if students can detect and avoid research groups characterized by disturbing practices. However, individual trainees who experience abusive relationships with a mentor may discover only too late that the practices that constitute the abuse were well known but were not disclosed to new initiates.

It is common practice for a graduate student to be supervised not only by an individual mentor but also by a committee that represents the graduate department or research field of the student. However, departmental oversight is rare for the postdoctoral research fellow. In order to foster good mentorship practices for all research trainees, many groups and institutions have taken steps to clarify the nature of individual and institutional responsibilities in the mentor–trainee relationship.[27]

FINDINGS AND CONCLUSIONS

The self-regulatory system that characterizes the research process has evolved from a diverse set of principles, traditions, standards, and customs transmitted from senior scientists, research directors, and department chairs to younger scientists by example, discussion, and informal education. The principles of honesty, collegiality, respect for others, and commitment to dissemination, critical evaluation, and rigorous training are characteristic of all the sciences. Methods and techniques of experimentation, styles of communicating findings, the relationship between theory and experimentation, and laboratory groupings for research and for training vary with the particular scientific disciplines. Within those disciplines, practices combine the general with the specific. Ideally, research practices reflect the values of the wider research community and also embody the practical skills needed to conduct scientific research.

Practicing scientists are guided by the principles of science and the standard practices of their particular scientific discipline as well as their personal moral principles. But conflicts are inherent among these principles. For example, loyalty to one's group of colleagues can be in conflict with the need to correct or report an abuse of scientific practice on the part of a member of that group.

Because scientists and the achievements of science have earned the respect of society at large, the behavior of scientists must accord not only with the expectations of scientific colleagues, but also with those of a larger community. As science becomes more closely linked to economic and political objectives, the processes by which scientists formulate and adhere to responsible research practices will be subject to increasing public scrutiny. This is one reason for scientists and research institutions to clarify and strengthen the methods by which they foster responsible research practices.

Accordingly, the panel emphasizes the following conclusions:

- The panel believes that the existing self-regulatory system in science is sound. But modifications are necessary to foster integrity in a changing research environment, to handle cases of misconduct in science, and to discourage questionable research practices.
- Individual scientists have a fundamental responsibility to ensure that their results are reproducible, that their research is reported thoroughly enough so that results are reproducible, and that significant errors are corrected when they are recognized. Editors of scientific journals share these last two responsibilities.
- Research mentors, laboratory directors, department heads, and senior faculty are responsible for defining, explaining, exemplifying, and requiring adherence to the value systems of their institutions. The neglect of sound training in a mentor's laboratory will over time compromise the integrity of the research process.
- Administrative officials within the research institution also bear responsibility for ensuring that good scientific practices are observed in units of appropriate jurisdiction and that balanced reward systems appropriately recognize research quality, integrity, teaching, and mentorship. Adherence to scientific principles and disciplinary standards is at the root of a vital and productive research environment.

• At present, scientific principles are passed on to trainees primarily by example and discussion, including training in customary practices. Most research institutions do not have explicit programs of instruction and discussion to foster responsible research practices, but the communication of values and traditions is critical to fostering responsible research practices and detering misconduct in science.

• Efforts to foster responsible research practices in areas such as data handling, communication and publication, and research training and mentorship deserve encouragement by the entire research community. Problems have also developed in these areas that require explicit attention and correction by scientists and their institutions. If not properly resolved, these problems may weaken the integrity of the research process.

NOTES

1. See, for example, Kuyper (1991).

2. See, for example, the proposal by Pigman and Carmichael (1950).

3. See, for example, Holton (1988) and Ravetz (1971).

4. Several excellent books on experimental design and statistical methods are available. See, for example, Wilson (1952) and Beveridge (1957).

5. For a somewhat dated review of codes of ethics adopted by the scientific and engineering societies, see Chalk et al. (1981).

6. The discussion in this section is derived from Mark Frankel's background paper, "Professional Societies and Responsible Research Conduct," included in Volume II of this report.

7. For a broader discussion on this point, see Zuckerman (1977).

8. For a full discussion of the roles of scientific societies in fostering responsible research practices, see the background paper prepared by Mark Frankel, "Professional Societies and Responsible Research Conduct," in Volume II of this report.

9. Selected examples of academic research conduct policies and guidelines are included in Volume II of this report.

10. See, for example, Holton's response to the criticisms of Millikan in Chapter 12 of *Thematic Origins of Scientific Thought* (Holton, 1988). See also Holton (1978).

11. See, for example, responses to the *Proceedings of the National Academy of Sciences* action against Friedman: Hamilton (1990) and Abelson et al. (1990). See also the discussion in Bailar et al. (1990).

12. Much of the discussion in this section is derived from a background paper, "Reflections on the Current State of Data and Reagent Exchange Among Biomedical Researchers," prepared by Robert Weinberg and included in Volume II of this report.

13. See, for example, Culliton (1990) and Bradshaw et al. (1990). For the impact of the inability to provide corroborating data or witnesses, also see Ross et al. (1989).

14. See, for example, Rennie (1989) and Cassidy and Shamoo (1989).

15. See, for example, the discussion on random data audits in Institute of Medicine (1989a), pp. 26-27.

16. For a full discussion of the practices and policies that govern authorship in the biological sciences, see Bailar et al. (1990).

17. Note that these general guidelines exclude the provision of reagents or facilities or the supervision of research as a criteria of authorship.

18. A full discussion of problematic practices in authorship is included in Bailar et al. (1990). A controversial review of the responsibilities of co-authors is presented by Stewart and Feder (1987).

19. In the past, scientific papers often included a special note by a named researcher, not a co-author of the paper, who described, for example, a particular substance or procedure in a footnote or appendix. This practice seems to have been abandoned for reasons that are not well understood.

20. Martin et al. (1969), as cited in Sigma Xi (1986), p. 41.

21. Huth (1988) suggests a "notice of fraud or notice of suspected fraud" issued by the journal editor to call attention to the controversy (p. 38). Angell (1983) advocates closer coordination between institutions and editors when institutions have ascertained misconduct.

22. Such facilities include Cambridge Crystallographic Data Base, GenBank at Los Alamos National Laboratory, the American Type Culture Collection, and the Protein Data Bank at Brookhaven National Laboratory. Deposition is important for data that cannot be directly printed because of large volume.

23. For more complete discussions of peer review in the wider context, see, for example, Cole et al. (1977) and Chubin and Hackett (1990).

24. The strength of theories as sources of the formulation of scientific laws and predictive power varies among different fields of science. For example, theories derived from observations in the field of evolutionary biology lack a great deal of predictive power. The role of chance in mutation and natural selection is great, and the future directions that evolution may take are essentially impossible to predict. Theory has enormous power for clarifying understanding of how evolution has occurred and for making sense of detailed data, but its predictive power in this field is very limited. See, for example, Mayr (1982, 1988).

25. Much of the discussion on mentorship is derived from a background paper prepared for the panel by David Guston. A copy of the full paper, "Mentorship and the Research Training Experience," is included in Volume II of this report.

26. Although the time to the doctorate is increasing, there is some evidence that the magnitude of the increase may be affected by the organization of the cohort chosen for study. In the humanities, the increased time to the doctorate is not as large if one chooses as an organizational base the year in which the baccalaureate was received by Ph.D. recipients, rather than the year in which the Ph.D. was completed; see Bowen et al. (1991).

27. Some universities have written guidelines for the supervision or mentorship of trainees as part of their institutional research policy guidelines (see, for example, the guidelines adopted by Harvard University and the University of Michigan that are included in Volume II of this report). Other groups or institutions have written "guidelines" (IOM, 1989a; NIH, 1990), "checklists" (CGS, 1990a), and statements of "areas of concern" and suggested "devices" (CGS, 1990c).

The guidelines often affirm the need for regular, personal interaction between the mentor and the trainee. They indicate that mentors may need to limit the size of their laboratories so that they are able to interact directly and frequently with all of their trainees. Although there are many ways to ensure responsible mentorship, methods that provide continuous feedback, whether through formal or informal mechanisms, are apt to be the most successful (CGS, 1990a). Departmental mentorship awards (comparable to teaching or research prizes) can recognize, encourage, and enhance the mentoring relationship. For other discussions on mentorship, see the paper by David Guston in Volume II of this report.

One group convened by the Institute of Medicine has suggested "that the university has a responsibility to ensure that the size of a research unit does not outstrip the mentor's ability to maintain adequate supervision" (IOM, 1989a, p. 85). Others have noted that although it may be desirable to limit the number of trainees assigned to a senior investigator, there is insufficient information at this time to suggest that numbers alone significantly affect the quality of research supervision (IOM, 1989a, p. 33).

3

Contemporary Research Environment

THE U.S. SCIENTIFIC RESEARCH ENTERPRISE

Brief Historical Perspective

The U.S. academic research enterprise of the 1990s differs in kind and scale from that of earlier decades. Once an informal, intimate, and paternalistic endeavor, research today is conducted as part of a more formal, complex, highly diversified enterprise that engages the talents of a broad spectrum of individuals and institutions. The organizational structures within which research is supported and performed, the climate within which research is conducted, and the criteria that define scientific achievement today are quite different from those that were in place previously.

The uniquely American multipurpose university was firmly established in the United States by 1890 and thereafter was gradually augmented by professional schools and institutes of technology (Geiger, 1990). Because institutional support for scientific work was scarce in the early part of the nineteenth century, research was usually an avocation rather than a profession. Later in the century, as the university system evolved and the idea of the pursuit of science for its own sake gained support, research was pursued as a full-time vocation (Daniels, 1967).

Throughout the first half of the twentieth century, universities retained the tradition of a community of independent scholars characterized by autonomy, individuality, and a diversity of research interests. Some faculty research was commercially or industrially oriented, particularly in the engineering schools and in chemistry departments. Some faculty followed government research interests in agriculture. Still others pursued independent research interests with small amounts of philanthropic support.

In response to the vital contributions of science and technology to U.S. victories in World War II, Bush (1945) and Steelman (1947) called for increased government support of research. The Bush report inspired a postwar relationship between government and the scientific community that sought to extend the successes of both government-organized projects such as the Manhattan Project and university-based research such as that performed at the Radiation Laboratory at the Massachusetts Institute of Technology. Both models of scientific work were eventually implemented, and the Bush report provided the blueprint for continued federal support of academic science through a decentralized process driven by investigator-initiated research proposals, eventually institutionalized with the establishment of the National Science Foundation in 1950.

The post-World War II years were thus the formative period for a more intimate relationship between the U.S. government and the scientific research community. The development of the contemporary system of federal support for university-based basic research and the education of new researchers provided the platform for the current preeminence of U.S. research (GUIRR, 1989). This system grew rapidly through the late 1960s (Brooks, 1989).

In the 1970s economic stagnation and concern about the cost of research and the social impact of science-based technologies led to a reexamination of the basic rationale governing federal investments in scientific research (GUIRR, 1989; Brooks, 1989). This reexamination led in turn to increased oversight and involvement of public officials with both science and technology. New regulatory requirements and new standards of accountability were imposed (OTA, 1986a).

In the 1980s renewed growth in federal funding for scientific research stimulated changes in the academic research environment. Support increased for research and development centers, large projects, and single-disease or single-technology programs, often called directed or mission-oriented research. But the accompanying increases in the size of academic administrative staffs and the amount of research overhead costs created concerns among sponsors and faculty.[1] In the face of increasing federal budget deficits in the late 1980s and decreasing economic and educational performance by the nation, motivations for funding research have focused increasingly on technological innovation, economic competitiveness, and education. This shifting rationale for federal support has been accompanied by demands for tighter management and oversight of research.

Current Concerns

How the contemporary research environment affects the integrity of the research process and the incidence of misconduct in science is poorly understood. But individual scientists and public officials have expressed concern about several factors that may foster dishonest behavior, which can range from subtle exaggeration of the value of research results to actual fabrication or falsification of research findings. One such factor is the pressure associated with producing research results to attract and maintain stable funding in a research system that cannot support all meritorious research proposals. Such pressure could erode the high standards of honesty and open collaboration that have traditionally characterized the scientific community. This and other concerns, coupled with heightened public awareness of waste, fraud, and abuse in other publicly supported activities, suggest that government oversight of the conduct of scientific research is likely to continue, if not increase. Such scrutiny has profound implications for the system of internal checks and balances in the research enterprise, which were designed for a research environment far removed from the forces of the political process.

THE CHANGING RESEARCH SCENE

Many factors have contributed to the evolving research scene, including the increasing complexity of contemporary research problems and instrumentation, the increasing costs of scientific research, changes in the rationale for supporting and monitoring government-funded research, and increased regulation of federal research. Other principal factors affecting the research environment include the scale, scope, and organization of research centers and groups; the changing character of collaborative efforts; the growing number of contenders for research funds; the reward system; and increasing emphasis on commercialization of research results. Combined, these factors exacerbate conflicts that have always been apparent to some extent in scientific research.

One example of the recent and profound changes characterizing the contemporary research environment is the changing nature of its basic organizational unit. Twenty years ago, a hypothetical laboratory group consisted of less than a half-dozen members. The group was small, closely knit, and composed of individuals who generally shared a common cultural heritage. The group accepted, often without conflict, a hierarchical structure of relationships and shared a common set of craft skills and moral standards, and its members followed well-understood lines of communication.

Today, although many research groups still consist of less than a dozen members, larger and more diverse research groups are becoming more common. The group members in large research teams differ in status; they include research investigators, undergraduate students, postdoctoral researchers, visiting faculty, and technicians. These individuals report, sometimes in an ill-defined manner, to a research director who frequently has many more professional and institutional obligations than his or her counterpart of 20 years ago.

Interpersonal conflicts and professional rivalries have always been part of the scientific culture. Yet good communication, good mentoring, and research supervision may be more difficult to achieve and to sustain in a large, complex, and democratic group environment (Phillip, 1991). Most research supervisors recognize the importance of good manners, civility, professional support, and personal interaction in their laboratories. However, the diverse social environment and the conflicting expectations of researchers offer increased opportunities for misunderstandings and unresolved disputes. If such disputes are not responsibly addressed, they sometimes can lead to allegations of misconduct in science, perhaps accompanied by an accusation that there has been a threat of reprisal. In the current environment, what has traditionally been regarded as an internal concern of a research laboratory or university can be escalated, sometimes rapidly, to a problem involving complex relationships and formal procedures between government agencies and research institutions. Questionable behavior in the research environment today is being publicized and publicly criticized.

Misconduct in science can occur, and allegations of misconduct must be treated seriously. But some complaints may simply reflect a poor research environment rather than actual misconduct in science. The best way to avoid or minimize research disputes is to establish a proper research environment. Research supervisors must devote attention to maintaining an atmosphere of open communication and cooperation in their research groups, with opportunity for appropriate participation by and recognition of all parties. Considering human relationships and interactions is an important aspect of good research practice.

Increased Size and Scope of the Research Enterprise

The U.S. research enterprise is larger today than it has ever been, whether measured in terms of numbers of institutions, research groups, investigators, postdoctoral fellows, technicians, graduate students, proposals, funding, research findings, articles, or knowledge produced. Scientific discoveries, patents, and publications in the post-World War II era all demonstrate the remarkable growth that has occurred in every field and discipline.

Federal funding of academic research and development has grown dramatically over the past 30 years, from less than $2 billion in 1958 to more than $8 billion in 1989, in constant 1988 dollars (GUIRR, 1989). The total number of scientists and engineers employed by universities increased from 120,000 in 1958 to 330,000 in 1988, while the number of Ph.D. degrees awarded annually increased from 6,000 to around 19,000 in the same period (GUIRR, 1989). Also during this time period, the annual growth rate in postdoctoral positions was 5 percent for science and 8 percent for engineering (NSB, 1989). While the number of investigators has been increasing, the number of investigator-initiated research proposals has been increasing at an even faster rate. The increase in proposals per investigator is probably related to the strategy of submitting multiple proposals to increase the likelihood of funding.[2]

The number of science and engineering articles published by academic scientists and engineers has doubled since 1965, with continued rapid growth through 1980 (NSF, 1990c). As with the number of proposals, the increase in articles has resulted from an increase both in the number of researchers and the number of articles produced per researcher per year. Similarly, the number of patents issued to universities, a different measure of research activity, has grown rapidly during the past decade.

Not only have the numbers of journals and articles increased, but the number of authors per article has also increased in fields such as high-energy physics, molecular biology, and clinical medicine. Huth, for example, reported that the mean number of authors per paper for the journals *Annals of Internal Medicine* and the *New England Journal of Medicine* rose from 1 in 1925 to 6 in 1985 (Huth, 1988).[3] He added that "in some papers the number of authors is clearly excessive for the intellectual activities represented" and that "the climbing number of authors per paper is tending to cheapen the value of authorship" (p. 40).

With more manuscripts submitted for publication and more proposals submitted for funding, the overall work load associated with critical evaluation has increased. There are concerns that peer review no longer operates as well as in earlier times, although the effects of increased volume on the operation of the system are not known (Chubin and Hackett, 1990).

Complexity of Collaboration

The increased emphasis on collaborative research is another indicator of change in the research environment. Before World War II, for example, scientific papers signed by more than four authors were practically nonexistent. Also extremely rare were papers that reported the results of collaborative efforts involving more than one laboratory or research team. But modern advances in the speed of travel and communication and in research instrumentation have changed the nature of scientific collaboration. Today, many important research papers involve collaboration among three or more laboratories, with a dozen or more authors in all. It is not unusual for authors or contributing laboratories to reside in more than one country. Although the senior investigators in these efforts may know each other personally, it is unlikely that the junior collaborators have ever met.

Different research groups may have different kinds of specialized skills, and complementary expert skills are likely to be the basis of the scientific collaboration. This type of interaction is very different, however, from earlier scientific exchanges in which all members of a research team shared the same laboratory environment and saw each other constantly during their work together.

Many of the achievements of modern science—of molecular biology, for example—have resulted from complex collaborative exchanges. Scientific advances in this field and others show that specialized collaborations can work effectively and are often indispensable to advancing knowledge. Nevertheless, the complexity of such operations, and the fact that many of the participants have limited personal interactions as well as limited abilities to evaluate the qualifications of others with different kinds of expertise, can give rise on occasion to conflicts and serious misunderstandings and can limit the effective operation of internal checks and balances.

Organization, Goals, and Management of Research Groups

Universities are characterized by decentralized organizational structures. The faculty traditionally govern academic programs. The faculty, in turn, are governed by a broad set of administrative and regulatory policies that affect the scientific research environment, increasingly so today. These policies reflect broad social concerns (e.g., about sexual harassment and equal opportunity) as well as matters explicitly related to the conduct of research (e.g., protection of human and animal subjects, regulation of toxic materials, and handling of hazardous equipment). In addition, many academic research institutions have now adopted policies regarding conflict of interest and the intellectual property developed by their employees.

Research in disciplinary specialties has traditionally been organized in a specific academic department. But research in many fields is now characterized by interdisciplinary approaches and is frequently carried out by individual academic investigators who, though they may have a departmental affiliation, are attached to independent research centers. Centers may be organized around common research interests (e.g., poverty, energy, the environment) or research styles and resource needs (e.g., surveys, computer modeling, synchrotron light sources). Center directors often assume responsibility for generating support, including ongoing support for facilities and core staff.

Research goals are increasingly linked, by sponsors and investigators, to specific social needs. Indeed, economic development has received explicit emphasis in recent years in some federal and most state-supported research. Research projects aimed at environmental, health, and other particular social problems have, since the 1960s, increasingly been carried out by interdisciplinary academic groups and research centers. Industry has often participated in and sponsored such activities and has provided a diversified source of funding. Research investigators in such organizations include tenured and junior faculty members, visiting scientists, nonfaculty research scientists, postdoctoral research fellows, graduate students, and technicians.

As a result of these trends, scientific research organizations today need an unaccustomed level of structure and efficient management to perform effectively. Many large research groups do not have organizational procedures to support the necessary level of management and oversight. Such circumstances can inhibit the effective resolution of disputes and even incidents of misconduct.

Issues related to authorship, allocation of credit, and data management practices often arise in large research groups. Teams of 100 Ph.D.s are common where research is dependent on major instruments. As instrumentation becomes more specialized, the team size, too, will grow, to 600 or more Ph.D.s in some instances. Some research team efforts are tightly coordinated, whereas other "big science" projects have a highly decentralized research culture. For example, the War on Cancer and the Human Genome Project have been described as combinations of "little science" initiatives. However, they typically follow a structured plan to achieve selected research objectives.

Research groups are governed by various management practices. Some groups operate in a collaborative style, choosing research problems through consultation among senior and junior investigators about the appropriate course to follow in pursuing interesting observations. Other groups adopt a more hierarchical style, whereby the principal investigator establishes a course of action for the research team as a whole and encourages efforts that contribute to the central mission of the director. In a few laboratories, research directors may discourage collegial discussion of new results or interpretation of findings or may foster competitive practices by assigning junior researchers to identical research problems.

Regulation and Accountability

Scientific research is increasingly subject to government regulations and guidelines that impose financial and administrative requirements and affect specific elements of the research process as well. Among the subjects of current research regulations are the assurance of a drug-free workplace, laboratory safety, proper use of human subjects and care of animal subjects, and care in the use of recombinant DNA and in the use of toxic and radioactive materials (OTA, 1986a). Regulatory requirements of the Public Health Service, the National Science Foundation, and the Department of Veterans' Affairs have also prompted, and in some cases required, research institutions to adopt policies and formal procedures to handle allegations of misconduct in science.

To assure the full compliance of investigators and institutions with these regulatory requirements, universities have expanded administrative and oversight functions. The associated costs in time and money have escalated tensions between administrators and faculties that would prefer to see the funds going into research. This is one of several issues that has caused schisms in the academic community.

Reward System

The criteria used to appoint, evaluate, and promote individual faculty members deeply influence the research enterprise. The rewards of a successful academic career traditionally include the personal gratification derived from scholarship and discovery, recognition by peers, and academic promotion and tenure, as well as enhanced responsibility and outside financial opportunities. The successful researcher can attract continuing research support and can enjoy a reputation that opens new opportunities for prestigious appointments.

The academic reward system today is influenced largely by research performance and productivity, sometimes measured by the number of publications or total amount of research support acquired by individual faculty. Intellectual contributions, teaching, and service to the university and the public are considered in varying degrees depending on the institution and discipline. However, there appears to be an imbalance, with emphasis on publication output and research support as the basis for promotion and tenure (Boyer, 1990).[4]

Quantitative measures of productivity have occasionally become substitutes for the critical evaluation of scientific work. This reliance on numbers arises in part because departmental peers are less able to evaluate the contribution of an individual researcher to large scientific projects or to interdisciplinary teams with an applied research approach. Attribution of credit among individuals on multiauthored publications is also difficult. Even when the form of an individual's contribution is clear, the significance of the contribution is often arguable.

The "publish-or-perish" dictum can lead to overspecialization, overemphasis on short-term projects, and the organization of research communication around the "least publishable unit." Theoretical approaches, including computer simulations, that yield especially rapid results can be favored over tedious programs of fundamental experiments. An excessive emphasis on quantitative measures of scientific productivity can penalize scientists who make responsible attempts to protect the quality of science (i.e., by delaying publication until they have completed a series of experiments instead of publishing each experiment). As Jackson and Prados (1983) have observed (p. 464):

> Good scientists may publish a lot or a little. But there is a very definite evil in a university that allows or encourages tenure committees to set standards of, say twenty published papers or abstracts in four years as a minimum requirement for consideration, or to discard as irrelevant any paper in a branch of science other than the tenure candidate's principal field of specialization.

Some institutions have responded to the emphasis on large numbers by limiting the number of publications reviewed for promotion and by rewarding nonresearch scholarship such as teaching and communicating science to a general audience (Kennedy, 1991; Harvard University Faculty of Medicine, 1988).

University–Industry Cooperation

The importance of contributing to economic development as a national research goal during the last decade has led to an emphasis on prompt transfer of fundamental research findings into commercial use. U.S. universities have often produced discoveries with practical significance, an achievement that has attracted the interest of both U.S. and foreign firms. In many areas of technological significance—microelectronics, biotechnology, materials science, instrumentation, and catalysis, for example—the interval between laboratory discovery and practical application has decreased. Rapid commercialization has provided increased incentives for joint industry-university research programs. Public desires to strengthen the competitive performance of U.S. industry have fostered academic research programs aimed at improving U.S. manufacturing.

A number of federal and state programs now encourage or require cooperation between universities, industry, and national laboratories. University-industry partnerships are implemented by a variety of mechanisms, including long-term agreements with one or more university research groups to pursue a subject of mutual interest, participation in research consortia, research contracts with specific program objectives, and informal collaborations. Consortia efforts, in which several companies combine with research groups at one or more universities to pursue a common research program, are another mechanism. Federal technology transfer programs, such as those in the Department of Defense and several Department of Energy programs at national laboratories, are other examples.

University–industry partnerships stimulate new ideas and innovation by both communities and motivate research teams to achieve important innovations of commercial value. But commercial relationships may introduce conflicts for academic investigators and the university. Some conflicts result from tension between the traditions of openness in the university, where prompt publication and free access to research results is required, and desires to restrict access to research results of proprietary value. Other conflicts can arise because personal profit and commercial interests can become explicit goals of individuals and institutions.

Despite efforts to minimize conflicts, there is growing concern in the scientific research community about the consequences of academic-commercial collaboration, especially in the area of clinical research. Many universities are adopting new and more stringent rules to govern conflict of interest and ownership of intellectual property, including categories of activities with differing requirements for disclosure as well as prohibited activities and relationships. An instructive case is the debate accompanying the adoption in 1990 of conflict-of-interest rules at the Harvard University Medical School (Harvard University Faculty of Medicine, 1990).

Conflicts of interest have the potential to affect peer review, publication and data management practices, training and mentorship, and other practices and behavior. For this reason, some scientific journals require authors to disclose sources of support and potential sources of bias when submitting their research papers.[5] Such conflicts also can influence the investigation of allegations of misconduct in science, especially if biases are not detected in the formation of investigatory panels that review and adjudicate misconduct complaints.

Although the panel does not believe that industry-university research arrangements present unique risks for misconduct in science, the self-serving interests associated with such arrangements pose issues that require institutional attention and oversight to ensure the integrity of the research process.

FINDINGS AND CONCLUSIONS

The contemporary research enterprise is far removed from that of the pre-World War II era. In particular, the academic research community, governed by traditions derived from an earlier model of a community of independent scholars who participated equally in academic governance, is challenged by the complexity of today's issues and of the environment in which research is conducted. Still, basic research continues to flourish, and faculty, postdoctoral fellows, and graduate students continue to contribute extraordinary research capability to science.

Concerns are apparent, however, and it is clear that key environmental factors require attention to protect the high standards of research integrity traditionally associated with scientists and their institutions. In reviewing changes within the scientific research enterprise, the panel reached the following conclusions:

• Scientific research is part of a larger and more complicated enterprise today, creating a greater need for individual and institutional attention to matters that affect the integrity of the research process. Scientists themselves and research institutions will be expected to play a more active role in ensuring that the activities performed by researchers are within the governance mechanisms of their institutions. The need for more explicit forms of institutional accountability and oversight is one price of the growth and success of the academic research enterprise.

• The growth and diversity of modern research call for institutions to accept explicit responsibility for fostering the integrity of the research process and for handling allegations of misconduct. In encouraging this acceptance, the panel is not suggesting that institutions assume responsibility for the correctness and accuracy of research results reported by their scientists or students. However, in recognizing that their faculty and research staff are responsible for maintaining the integrity of the research process, institutions should retain and accept certain explicit obligations. Principal among these is providing a research environment that fosters honesty, integrity, and a sense of community. Institutions should strive to attain a research enterprise that emphasizes and rewards excellence in science, quality rather than quantity, openness rather than secrecy, and collegial obligations rather than opportunistic behavior in appointment, promotion, tenure, and other career decisions. Research institutions should also recognize the risks that are inherent in self-regulation and strive to involve outside parties, when appropriate, in investigating or evaluating the conduct of their own members. Steps toward achieving these goals are discussed in Chapter 6.

• The increased size, specialization, and diversity of research groups, and other changes in the social relationships of their members, have stimulated personal conflicts and misunderstandings, including disputes within and between research groups about fairness and allocation of credit. These disputes may be prevented by positive efforts to foster responsible research practices and by taking preemptive actions, such as prior discussion and agreement on allocation of credit, to promote a harmonious work environment that encourages collegiality, collaboration, and productivity. Frank discussions, both formal and informal, possibly aided by outside mediators, are additional tools to use in addressing these disputes.

• The issues associated with conflict of interest in the academic research environment are sufficiently problematic that they deserve thorough study and analysis by major academic and scientific organizations, including the National Academy of Sciences. Disclosure, either public or institutional, is essential to controlling conflict of interest, and some universities and scientific journals prohibit certain forms of commercial contractual arrangements by their members or authors. But the responsibility for such disclosure rests with scientists themselves.

• The research environment is stressful and yet conducive to the remarkable productivity of researchers. The rewards for successful research are greater now than in the past, but today's rapid pace of development may undermine critical internal checks and balances and may increase opportunities for misrepresentation or distortion of research results. Thus the scientific community must organize to reinforce its standards and to ensure the responsible conduct of research.

NOTES

1. See, for example, Association of American Universities (1988).

2. The Office of Technology Assessment suggests that a "kind of lottery mentality appears to have taken hold in the 1980s: the more grant proposals submitted, the greater the probability that one would be funded" (OTA, 1990, p. 10).

3. The mean is represented by rounding off to one significant figure.

4. See also, for example, Angell (1986).

5. See, for example, the editorial policies of the *New England Journal of Medicine* (1992).

4

Misconduct in Science—
Incidence and Significance

Estimates reported in government summaries, research studies, and anecdotal accounts of cases of confirmed misconduct in science in the United States range between 40 and 100 cases during the period from 1980 to 1990.[1] The range reflects differences in the definitions of misconduct in science, uncertainties about the basis for "confirmed" cases, the time lag between the occurrence and disclosure of some cases, and potential overlap between government summaries (which are anonymous) and cases identified by name in the research literature.

When measured against the denominator of the number of research awards or research investigators, the range of misconduct-in-science cases cited above is small.[2] Furthermore, less than half of the allegations of misconduct received by government agencies have resulted in confirmed findings of misconduct in science. For example, after examining 174 case files of misconduct in science in the period from March 1989 through March 1991, the Office of Scientific Integrity in the Public Health Service found evidence of misconduct in fewer than 20 cases, although 56 investigations, mostly conducted by universities, were still under way (Wheeler, 1991).

However, even infrequent incidents of misconduct in science raise serious questions among scientists, research sponsors, and the public about the integrity of the research process and the stewardship of federal research funds.

INCIDENCE OF MISCONDUCT IN SCIENCE— PUBLISHED EVIDENCE AND INFORMATION

The incidence of misconduct in science and the significance of several confirmed cases have been topics of extensive discussion. Measures of the incidence of misconduct in science include (1) the number of allegations and confirmations of misconduct-in-science cases recorded and reviewed by government agencies and research institutions and (2) data and information presented in analyses, surveys, other studies, and anecdotal reports.

Some observers have suggested that incidents of misconduct in science are underreported. It may be difficult for co-workers and junior scientists, for example, graduate students and postdoctoral fellows, to make allegations of misconduct in science because of lack of supporting evidence and/or fear of retribution. The significant professional discrimination and economic loss experienced by whistleblowers as a result of reporting misconduct are well known and may deter others from disclosing wrongdoing in the research environment.

Government Statistics on Misconduct in Science

Owing to differing perspectives on the role of government and research institutions in addressing misconduct in science, and to discrepancies in the number of allegations received by government offices, the number of open cases, and the cases of misconduct in science confirmed by research institutions or government agencies, many questions remain to be answered. These areas of uncertainty and disagreement inhibit the resolution of issues such as identifying the specific practices that fit legal definitions of misconduct in science; agreeing on standards for the evidence necessary to substantiate a finding of misconduct in science; clarifying the extent to which investigating panels can or should consider the intentions of the accused person in reaching a finding of misconduct in science; assessing the ability of research institutions and government agencies to discharge their responsibilities effectively and handle misconduct investigations appropriately; determining the frequency with which misconduct occurs; achieving consensus on the penalties that are likely to be imposed by diverse institutions for similar types of offenses; and evaluating the utility of allocating substantial amounts of public and private resources to handle allegations, only a few of which may result in confirmed findings of misconduct. The absence of publicly available summaries of the investigation and adjudication of incidents of misconduct in science inhibits scholarly efforts to examine how prevalent misconduct in science is and to evaluate the effectiveness of governmental and institutional treatment and prevention programs.

As a result, analyses of and policies related to misconduct in science are often influenced by information derived from a small number of cases that have received extensive publicity. The panel has not seen evidence that would help determine whether these highly publicized cases are representative of the broader sample of allegations or confirmed incidents of misconduct in science. One trend should be emphasized, however. The highly publicized cases often involve charges of falsification and fabrication of data, but the large majority of cases of confirmed misconduct in science have involved plagiarism (NSF, 1991a; Wheeler, 1991). Possible explanations for this trend are that plagiarism is more clearly identifiable by the complainants and more easily proved by those who investigate the complaint.

Five semiannual reports prepared by the National Science Foundation's Office of Inspector General (NSF 1989c; 1990a,b; 1991a,c) and a 1991 annual report prepared by the Office of Scientific Integrity Review of the Department of Health and Human Services (DHHS, 1991b) are the first systematic governmental efforts to analyze characteristics of a specific set of cases of misconduct in science. Although the treatment of some individual cases reported in these summaries has been the subject of debate and controversy, the panel commends these analyses as initial efforts and suggests that they receive professional review and revisions, if warranted.

National Science Foundation

The National Science Foundation's (NSF's) Office of Inspector General (OIG) received 41 allegations of misconduct in science in FY 1990 and reviewed another group of 6 allegations received by NSF prior to 1990 (NSF, 1990b).[3] From this group of 47 allegations, OIG closed 21 cases by the end of FY 1990. In three cases NSF made findings of misconduct in science; in another four cases, NSF accepted institutional findings of misconduct in science. NSF officials caution that, in their view, future cases may result in a larger percentage of confirmed findings of misconduct because many of the open cases raise complicated issues that require more time to resolve.[4]

The panel matched the 41 allegations reviewed by NSF in FY 1990 against the definitions of misconduct in science used by NSF at that time (Table 4.1).

The NSF's Office of the Director recommended the most serious

TABLE 4.1 Allegations of Misconduct in Science Reviewed in FY 1990 by the National Science Foundation

Category	Number of Allegations
Fabrication or falsification	9
Plagiarism	20
Other deviant research practices	8[a]
Violations of other research conduct regulations	1[b]
Violations of other legal requirements governing research	4[c]
TOTAL	41[d]

NOTE: The table represents the categories assigned by the panel to the allegations themselves. NSF's OIG does not necessarily endorse these categories, nor does it necessarily regard all these cases as exemplifying misconduct in science.

[a]Allegations of deviant practices included unauthorized use of research preparations, failure to identify original authors of proposal, tampering with others' experiments, discrimination by a reviewer or research investigator, and exploitation of a subordinate.

[b]Alleged violation of recombinant DNA regulations.

[c]Alleged violations included financial conflict of interests under award by an investigator or reviewer, NSF staff mishandling of proposal or award, and violation of a sanction against a principal investigator.

[d]Some allegations involved more than one form of misconduct.

SOURCE: Based on data from Office of Inspector General, National Science Foundation (personal communications on December 27, 1990, and February 22, 1991).

penalty (debarment for 5 years) in a case involving charges of repeated incidents of sexual harassment, sexual assault, and threats of professional and academic blackmail by a co-principal investigator on NSF-funded research (NSF, 1990b, p. 21). Following an investigation that involved extensive interviews and affidavits, NSF's OIG determined that "no federal criminal statutes were violated . . . [but that] the pattern and effect of the co-principal investigator's actions constituted a serious deviation from accepted research practices" (NSF, 1990b, p. 21). NSF's OIG further determined that these incidents were "an integral part of this individual's performance as a researcher and research mentor and represented a serious deviation from accepted research practices" (p. 27). However, reports of this particular case have caused some scientists to express concern that the scope of the definition of misconduct in science may be inappropriately broadened into areas designated by the panel as "other misconduct," such as sexual harassment.

Department of Health and Human Services

In FY 1989 and FY 1990, following the creation of the Office of Scientific Integrity (OSI), the Department of Health and Human Services (DHHS) received a total of 155 allegations of misconduct in science, many of which had been under review from earlier years by various offices within the Public Health Service (PHS).[5] In April 1991, OSI reported that since its formation it had closed about 110 cases, most of which did not result in findings of misconduct in science.

The Office of Scientific Integrity Review (OSIR), in the office of the assistant secretary for health, reviewed 21 reports of investigations of misconduct in science in the period from March 1989 to December 1990, some of which involved multiple charges.[6] The cases reviewed by OSIR had been forwarded to that office by OSI and had completed both an inquiry and investigation stage. Findings of misconduct in science, engaged in by 16 individuals, were made in 15 of the reports of investigations reviewed by OSIR. The OSIR's summary of findings is given in Table 4.2.

The OSIR recommended debarment in six cases, the most extreme administrative sanction available short of referral to the Justice Department for criminal prosecution. Actions to recover PHS grant funds were undertaken in two cases.

Consequences of Confirmed Misconduct

Confirmed findings of misconduct in science can result in governmental penalties, such as dismissal or debarment, whereby individuals or institutions can be prohibited from receiving government grants or contracts on a temporary or permanent basis (42 C.F.R. 50). An individual who presents false information to the government in any form, including a research proposal, employment application, research report, or publication, may be subject to prosecution under the False Claims Act (18 U.S.C. 1001). At least one case of criminal prosecution against a research scientist, for example, rested on evidence that the scientist had provided false research information in research proposals and progress reports to a sponsoring agency.[7] Similar prosecutions have occurred in connection with some pharmaceutical firms or contract laboratories that provided false test data in connection with licensing or government testing requirements (O'Reilly, 1990).

Government regulations on misconduct in science provide a separate mechanism through which individuals and institutions can be

TABLE 4.2 Findings of Misconduct in Science in Cases Reviewed by the Office of Scientific Integrity Review, Department of Health and Human Services, March 1989 to December 1990

Type of Allegation	Findings of Misconduct (15 investigations)
Fabrication or falsification	6
Plagiarism	5
Other deviant research practices	7
TOTAL	18[a]

[a]The total of findings of misconduct is larger than the number of investigations because some cases had multiple findings.

SOURCE: Department of Health and Human Services (1991b).

subjected to government penalties and criminal prosecution if they misrepresent information from research that is supported by federal funds, even if the information is not presented directly to government officials. Research institutions and scientific investigators who apply for and receive federal funds are thus expected to comply with high standards of honesty and integrity in the performance of their research activities.

Government Definitions of Misconduct in Science— Ambiguity in Categories

The PHS's misconduct-in-science regulations apply to research sponsored by all PHS agencies, including the National Institutes of Health, the Alcohol, Drug Abuse, and Mental Health Administration, the Centers for Disease Control, the Food and Drug Administration, and the Agency for Health Care Policy and Research. The PHS defines misconduct in science as "fabrication, falsification, plagiarism, or other practices that seriously deviate from those that are commonly accepted within the scientific community for proposing, conducting, or reporting research. It does not include honest error or honest differences in interpretations or judgments of data" (DHHS, 1989a, p. 32447).[8]

The PHS's definition does not further define fabrication, falsification, plagiarism, or other serious deviations from commonly accepted research practices. The ambiguous scope of this last category is a topic of major concern to the research community because of the perception that it could be applied inappropriately in cases of disputed scientific judgment.

The first annual report of the DHHS's OSIR suggests the types of alleged misconduct in science that might fall within the scope of this category (DHHS, 1991b):

• Misuse by a journal referee of privileged information contained in a manuscript,

• Fabrication of entries or misrepresentation of the publication status of manuscripts referenced in a research bibliography,

• Failure to perform research supported by a PHS grant while stating in progress reports that active progress has been made,

• Improper reporting of the status of subjects in clinical research (e.g., reporting the same subjects as controls in one study and as experimental subjects in another),

• Preparation and publication of a book chapter listing co-authors who were unaware of being named as co-authors,

• Selective reporting of primary data,

• Unauthorized use of data from another investigator's laboratory,

• Engaging in inappropriate authorship practices on a publication and failure to acknowledge that data used in a grant application were developed by another scientist, and

• Inappropriate data analysis and use of faulty statistical methodology.

The panel points out that most of the behaviors described above, such as the fabrication of bibliographic material or falsely reporting research progress, are behaviors that fall within the panel's definition of misconduct in science proposed in Chapter 1.

The NSF's definition (NSF, 1991b) is broader than that used by the PHS[9] and extends to nonresearch activities supported by the agency, such as science education. NSF also includes in its definition of misconduct in science acts of retaliation against any person who provides information about suspected misconduct and who has not acted in bad faith.

The panel believes that behaviors such as repeated incidents of sexual harassment, sexual assault, or professional intimidation should be regarded as other misconduct, not as misconduct in science, because these actions (1) do not require expert knowledge to resolve complaints and (2) should be governed by mechanisms that apply to all institutional members, not just those who receive government research awards. Practices such as inappropriate authorship, in the panel's view, should be regarded as questionable research practices, because they do not fit within the rationale for misconduct in science as defined by the panel in Chapter 1.

The investigation of questionable research practices as incidents of alleged misconduct in science, in the absence of consensus about the nature, acceptability, and damage that questionable practices cause, can do serious harm to individuals and to the research enterprise. Institutional or regulatory efforts to determine "correct" research methods or analytical practices, without sustained participation by the research community, could encourage orthodoxy and rigidity in research practice and cause scientists to avoid novel or unconventional research paradigms.[10]

Reports from Local Institutional Officials

Investigatory Reports

Government regulations currently require local institutions to notify the sponsoring agency if they intend to initiate an investigation of an allegation of misconduct in science. The institutions are also required to submit a report of the investigation when it is completed. These reports, in the aggregate, may provide a future source of evidence regarding the frequency with which misconduct-in-science cases are handled by local institutions.

Although some investigatory reports have been released on an ad hoc basis, research scientists generally do not have access to comprehensive summaries of the investigatory reports prepared or reviewed by government agencies. The absence of such summaries impedes informed analysis of misconduct in science and inhibits the exchange of information and experience among institutions about factors that can contribute to or prevent misconduct in science.

Other Institutional Reports

The perspectives and experiences of institutional officials in handling allegations of misconduct in science are likely in the future to be important sources of information about the incidence of misconduct. This body of experience is largely undocumented, and most institutions do not maintain accessible records on their misconduct cases because of concerns about individual privacy and confidentiality, as well as concerns about possible institutional embarrassment, loss of prestige, and lawsuits.

The DHHS's regulations now require grantee institutions to provide annual reports of aggregate information on allegations, inquiries, and investigations, along with annual assurances that the institutions have an appropriate administrative process for handling allegations of misconduct in science (DHHS, 1989a). The institutional reports filed in early 1991 were not available for this study. These institutional summaries could eventually provide an additional source of evidence regarding how frequently misconduct in science addressed at the local level involves biomedical or behavioral research. If the reports incorporate standard terms of reference, are prepared in a manner that facilitates analysis and interpretation, and are accessible to research scientists, they could provide a basis for making independent judgments about the effectiveness of research institutions in handling allegations of misconduct in science. The NSF's regulations do not require an annual report from grantee institutions.

International Studies

Cases of misconduct in science have been reported and confirmed in other countries. The editor of the *British Medical Journal* reported in 1988 that in the 1980s at least five cases of misconduct by scientists had been documented in Britain and five cases had been publicly disclosed in Australia (Lock, 1988b, 1990). As a result of a "nonsystematic" survey of British medical institutions, scientists, physicians, and editors of medical journals, Lock cited at least another 40 unreported cases.

There has been at least one prominent case of misconduct in science in India recently (Jayaraman, 1991). Several cases of misconduct in science and academic plagiarism have been recorded in Germany (Foelsing, 1984; Eisenhut, 1990).

Analyses, Surveys, and Other Reports

Hundreds of articles on misconduct in science have been published in the popular and scholarly literature over the past decade. The study panel's own working bibliography included over 1,100 such items.

Although highly publicized reports about individual misconduct cases have appeared with some frequency, systematic efforts to analyze data on cases of misconduct in science have not attracted significant interest or support within the research community until very recently. Research studies have been hampered by the absence of information and statistical data, lack of rigorous definitions of misconduct in science, the heterogeneous and decentralized nature of the research environment, the complexity of misconduct cases, and the confidential and increasingly litigious nature of misconduct cases (U.S. Congress, 1990b; AAAS-ABA, 1989).

As a result, only a small number of confirmed misconduct cases have been the subject of scholarly examination. The results of these studies are acknowledged by their authors to be subject to statistical bias; the sample, which is drawn primarily from public records, may or may not be representative of the larger pool of cases or allegations. Preliminary studies have focused primarily on questions of range, prevalence, incidence, and frequency of misconduct in science. There has been little effort to identify patterns of misconduct or questionable practices in science. Beyond speculation, very little is known about the etiology, dynamics, and consequences of misconduct in science. The relationship of misconduct in science to factors in the contemporary research environment, such as the size of research teams, financial incentives, or collaborative research efforts, has not been systematically evaluated and is not known.

Woolf Analysis

Patricia Woolf of Princeton University, a member of this panel, has analyzed incidents of alleged misconduct publicly reported from 1950 to 1987 (Woolf, 1981, 1986, 1988a).

Woolf examined 26 cases of misconduct identified as having occurred or been detected in the period from 1980 to 1987, the majority of which (22 cases) were in biomedical research. Her analysis indicated that 11 of the institutions associated with the 26 cases were prestigious schools and hospitals, ranked in the top 20 in the Cole and Lipton (1977) evaluation of reputation. Woolf found that a "notable percentage" of the individuals accused of misconduct were from highly regarded institutions: "seven graduated from the top twenty schools" (Woolf, 1988a, p. 79), as ranked by reputation, an important finding that deserves further analysis. She also suggested that because cases of misconduct are often handled locally, the total number of cases is likely to be larger than reported in the public record (Woolf, 1988a).

The types of alleged misconduct reported in the cases analyzed by Woolf, some of which involved more than one type, included plagiarism (4 cases); falsification, fabrication, and forgery of data (12 cases); and misrepresentation and other misconduct (12 cases). She suggested that "plagiarism is almost certainly under-represented in this survey, as it appears to be handled locally and without publicity whenever possible" (Woolf, 1988a, p. 83).

Woolf identified several important caveats, noted below, that still apply to all systematic efforts to analyze the characteristics and demography of misconduct in science (Woolf, 1988a, p. 76):

- *Small number of instances.* There are not enough publicly known cases to draw statistically sound conclusions or make significant generalizations, and those that are available are a biased sample of the population of actual cases.
- *Blurred categories.* It is not possible in all cases to cleanly separate misconduct in science from falsification in drug trials or laboratory tests. Similarly, one person may indulge in plagiarism, fabrication, *and* falsification.
- *Incomplete information.* Some information about reported instances is not yet available.
- *Variety of sources.* The sources of information (for Woolf's analysis) include public accounts, such as newspaper reports, as well as original documents and interview material. They are not all equally reliable with regard to dates and other minor details.
- *Unclear resolution.* Disputed cases that have nevertheless been "settled" are included (in Woolf's analysis). In some highly publicized cases of alleged misconduct in science, the accused scientist has *not* admitted, and may have specifically denied, misconduct in science.

OSIR Analysis

The DHHS's OSIR prepared a first annual report in early 1991 that analyzed data associated with investigations of misconduct in science reviewed by that office in the period March 1989 through December 1990 (DHHS, 1991b). The report examined misconduct investigations carried out by research institutions and by the OSI.

Seniority of Subjects of Misconduct Cases in Woolf and OSIR Analyses. Both Woolf and the OSIR examined the rank of individuals who have been the subjects of misconduct-in-science cases. Although some have speculated that junior scientists might be more likely to engage in misconduct in science, both Woolf's analysis and the OSIR's analysis suggest that misconduct in science "did not occur primarily among junior scientists or trainees" (DHHS, 1991b, p. 7). Their preliminary studies suggest that the incidence of misconduct is likely to be greater among senior scientists (Table 4.3), a finding that deserves further analysis.

Detection of Misconduct in Science in Woolf and OSIR Analyses. Woolf and the OSIR examined processes used to detect incidents of con-

TABLE 4.3 Academic Ranks of Subjects in Confirmed Cases of Misconduct in Science

Rank	Number of Subjects	
	1980-1987[a]	1989-1990[b]
Full or asssociate professor, or senior scientist/laboratory chief	13	7
Assistant professor	2	4
Research associate/fellow	3	3
Various posts held	5	na
No academic appointment/technicians	2	2
Unknown	1	na
	26	16

[a]Data from Woolf (1988a).
[b]Department of Health and Human Services (1991b).

firmed or suspected misconduct in science and also analyzed the status of individuals who disclosed these incidents (Tables 4.4 and 4.5). Their analyses indicate that existing channels within the peer review process and research institutions do provide information about misconduct in science. Initial reports were often made by supervisors, collaborators, or subordinates who were in direct contact with the individual suspected of misconduct. These findings contradict opinions that checks such as peer review, replication of research, and journal reviews do not help identify instances of misconduct.

However, the panel notes that supervisors, colleagues, and subordinate personnel may report misconduct in science at their peril. The honesty of individuals who hold positions of respect or prestige cannot be easily questioned. It can be particularly deleterious for junior or temporary personnel to make allegations of misconduct by their superiors. Students, research fellows, and technicians can jeopardize current positions, imperil progress on their research projects, and sacrifice future recommendations from their research supervisors by making allegations of misconduct by their co-workers.

The Acadia Institute Survey

One provocative study of university officers' experience with misconduct in science is a 1988 survey of 392 deans of graduate studies from institutions affiliated with the Council of Graduate Schools (CGS).[11] [12] The survey was conducted with support from NSF and the American

TABLE 4.4   Primary Sources of Detection of Alleged Misconduct (1980 to 1987)

Factor	Number of Cases
Admission	2
Co-worker or former co-worker reported:	
Laboratory suspicions, irregular procedures	13
Misuse of funds	1
Inability to replicate or continue work	8
Institutional review board raised questions	1
Scientists at other institutions reported	
suspicions (including inability to replicate work)	6
Editorial peer review	3
Promotion review of publications	1
Formal audit	1
Protest by original author (plagiarism)	3
Unknown	2

NOTE: Some instances were or seem to have been suspected or detected at about the same time by more than one factor. From the available record it is difficult to make a clear distinction between factors that enabled detection of misconduct in science and those used to demonstrate or prove it.

SOURCE: Data from Woolf (1988a).

Association for the Advancement of Science. Approximately 75 percent (294) of the graduate deans responded to the survey.

The Acadia Institute survey data indicate that 40 percent (118) of the responding graduate deans had received reports of *possible* faculty misconduct in science during the previous 5 years. Two percent (6) had received more than five reports. These figures suggest that graduate deans have a significant chance of becoming involved in handling an allegation of misconduct in science.

The survey shows that about 190 allegations of misconduct in science were addressed by CGS institutions over the 5-year period (1983 to 1988) reported in the survey. It is not known whether any or all of these allegations were separately submitted to government offices concerned with misconduct in science during this time period, although overlap is likely.

The Acadia Institute survey also suggests, not surprisingly, that allegations of misconduct in science are associated with institutions that receive significant amounts of external research funding. As noted in the NSF's OIG summary report of the Acadia Institute survey: "Of the institutions receiving more than $50 million in external research funding annually, 69 percent [36] had been notified of possi-

TABLE 4.5 Status of Individual Bringing Allegations

Status	Number of Cases
Supervisor (e.g., department chair, laboratory chief)	4
Colleague (scientific associate of about the same seniority or status)	4
Collaborator	4
Junior scientific associate	2
Graduate student or postdoctoral trainee	5
Laboratory technician	3
Chair of a department at another institution	1
Self (self-report of misconduct by the subject)	1

SOURCE: Department of Health and Human Services (1991b).

ble faculty misconduct. Among institutions receiving less than $5 million, only 19 percent [14] had been so notified" (NSF, 1990d, pp. 2-3).

When asked about cases of *verified* misconduct by their faculties during the previous 5 years, 20 percent (59) of all the responding graduate deans indicated such instances. Among universities with over $50 million per year in external funding (about 55 institutions fell within this category in 1988), 41 percent (20) had some verified misconduct, according to responses of graduate deans participating in the Acadia Institute survey. The actual number of cases associated with these percentages, which is small, is consistent with the panel's observation that the total number of confirmed cases of misconduct in science is very small. Nevertheless, reports indicating that prestigious research institutions consistently receive, and confirm, allegations of misconduct in science are disturbing.

Other Reports

Bechtel and Pearson. Bechtel and Pearson (1985) examined both the question of prevalence of misconduct in science and the concept of deviant behavior by scientists as part of a larger exploration of "elite occupational deviance" that included white collar crime. The authors reviewed 12 cases of misconduct in science, drawn from reports in the popular and scientific press in the 1970s and early 1980s. They found that available evidence was inadequate to support accurate generalizations about how widespread misconduct in science might be. As to the causes of deviant behavior, the authors concluded that "in the debate between those who favor individualistic explanations based on psychological notions of emotional disturbance, and the critics of big science who blame the increased pressures for promotion, tenure, and recognition through publications, one tends to see greater merit in the latter" (p. 244). They suggested that further systematic examination is required to determine the appropriate balance between individual and structural sources of deviant behavior.

Sigma Xi Study. As part of a broader survey it conducted in 1988, Sigma Xi, the honor society for scientific researchers in North America, asked its members to respond to the following statement: "Excluding gross stupidities and/or minor slip ups that can be charitably dismissed (but not condoned), I have direct knowledge of fraud (e.g., falsifying data, misreporting results, plagiarism) on the part of a professional scientist."[13]

Respondents were asked to rank their agreement or disagreement with the statement on a five-point scale. The survey was mailed to 9,998 members of the society; about 38 percent responded (which indicates a possible source of bias).

Although 19 percent of the Sigma Xi respondents indicated that they had direct knowledge of fraud by a scientist, it is not certain from the survey whether direct knowledge meant personal experience with or simply awareness of scientific fraud. It is also possible that some respondents were referring to identical cases, and respondents may have reported knowledge of cases gained secondhand. Furthermore, it is not clear what information can be gained by having respondents rank "direct knowledge" on a five-point scale of agreement and disagreement.

Additional Information. Estimates about the incidence of misconduct in science have ranged from editorial statements that the scientific literature is "99.9999 percent pure" to reader surveys published in scientific journals indicating that significant numbers of the respondents have had direct experience with misconduct of some sort in science.[14] The broad variance in these estimates has not resolved uncertainties about the frequency with which individuals or institutions actually encounter incidents of misconduct in science.

In March 1990, the NSF's OIG reported that, based on a comprehensive review of the results from past surveys that attempted to measure the incidence of misconduct in science, "the full extent of misconduct is not yet known" (NSF, 1990d, p. 9). The NSF reports found that only a few quantitative studies have examined the extent of misconduct in science and that prior survey efforts had poor response rates, asked substantively different questions, and employed varying definitions of misconduct. These efforts have not yielded a database that would provide an appropriate foundation for findings and conclusions about the extent of misconduct in science and engineering.[15]

FINDINGS AND CONCLUSIONS

The panel found that existing data are inadequate to draw accurate conclusions about the incidence of misconduct in science or of questionable research practices. **The panel points out that the number of confirmed cases of misconduct in science is low compared to the level of research activity in the United States. However, as with all forms of misconduct, underreporting may be significant; federal agencies have only recently imposed procedural and reporting requirements that may yield larger numbers of reported cases. The possibility of underreporting can neither be dismissed nor confirmed at this time. More research is necessary to determine the full extent of misconduct in science.**

Regardless of the incidence, the panel emphasizes that even infrequent cases of misconduct in science are serious matters. The number of confirmed incidents of misconduct in science, together with the possibility of underreporting and the results presented in some preliminary studies, indicate that misconduct in science is a problem that cannot be ignored. The consequences of even infrequent cases of misconduct in science require that attention be given to appropriate methods of treatment and prevention.

NOTES

1. Reports of cases involving findings of misconduct in science were provided to the panel by DHHS and NSF. These reports indicate a total of 15 cases of findings of misconduct in science by DHHS in the period from March 1989 to December 1990 and 3 cases of findings of misconduct in science by NSF in the period from July 1989 to September 1990. See NSF (1990b) and DHHS (1991b). Information was also provided in a personal communication from Donald Buzzelli, staff associate, OIG, NSF, February 1, 1991.

Congressional testimony by and telephone interviews with NIH and ADAMHA officials indicated that in the period from 1980 to 1987, roughly 17 misconduct cases handled by these agencies resulted in institutional findings of research misconduct, some of which are included in the Woolf analysis discussed below. During this same period, NSF made findings of misconduct in science in seven cases. See the testimony of Katherine Bick and Mary Miers in U.S. Congress (1989a); see also Woolf (1988a).

The report by Woolf (1988a) identified 40 publicly reported cases of alleged misconduct in science in the period from 1950 to 1987, many of which involved confirmed findings of misconduct. Another two dozen or so cases of alleged misconduct in science were reported in congressional hearings in the 1980s. Some of the cases discussed in congressional hearings and in the Woolf analysis are included in the NSF and DHHS reports mentioned above. Some cases discussed in congressional hearings are still open, and the remainder have been closed without an institutional finding of misconduct in science.

The estimate of confirmed cases of misconduct in science does not include cases in which research institutions have made findings of misconduct, unless these cases are included in the Woolf analysis or the congressional hearings mentioned above. During the time of this study, there were no central records for institutional reports on misconduct in science that would indicate the frequency with which these organizations found allegations to have merit.

Finally, several authors have reviewed selected cases of misconduct in science, both contemporary and historical. The most popular accounts are a book by Broad and Wade (1982), who cite 34 cases of "known or suspected cases of scientific fraud" ranging from "ancient Greece to the present day"; a book by Klotz (1985); and one by Kohn (1986), who cites 24 cases of "known or suspected misconduct." These texts, and the government reports, congressional hearings, and Woolf analysis cited above, discuss many of the same cases.

2. The preamble to the PHS's 1989 regulations for scientific misconduct notes that "reported instances of scientific misconduct appear to represent only a small fraction of the total number of research and research training awards funded by the PHS" (DHHS, 1989a, p. 32446). The preamble to the NSF's 1987 misconduct regulations states that "NSF has received relatively few allegations of misconduct or fraud occurring in NSF-supported research or . . . proposals" (NSF, 1987, p. 24466).

Furthermore, according to the National Library of Medicine, during the 10-year period from 1977 to 1986, about 2.8 million articles were published in the world's biomedical literature. The number of articles retracted because of the discovery of fraud or falsification of data was 41, less than 0.002 percent of the total. See Holton (1988), p. 457.

3. Analyses of the NSF's experience are complicated by the fact that different offices have held authority for handling research misconduct cases. Prior to the creation of the OIG in March 1989, this authority was assigned to the NSF's Division of Audit and Oversight. The OIG "inherited" approximately 19 case files, and it received 6 new allegations of research misconduct during FY 1989. NSF officials reported in 1987 that NSF had examined 12 charges of research misconduct, 7 of which were found to be warranted, of which 3 were considered minor violations. See Woolf (1988a).

4. Personal communication, OIG, NSF, February 1, 1991.

5. Personal communication, Jules Hallum, director, OSI, February 27, 1991.

6. Four of these investigations were conducted by the PHS. Sixteen were conducted by outside, primarily grantee, institutions. One additional investigation was an intramural case within the PHS.

7. See the documentation regarding the case of psychologist Stephen Breuning as detailed in the DHHS's Report and Recommendations of a Panel to Investigate Allegations of Scientific Misconduct under Grants MH-32206 and MH-37449, April 20, 1987.

8. The definition excludes violations of regulations that govern human or animal experimentation, financial or other record-keeping requirements, or the use of toxic or hazardous substances. It applies to individuals or institutions that apply for as well as those that receive extramural research, research-training, or research-related grants or cooperative agreements under the PHS, and to all intramural PHS research. In the proposed rule, the PHS's definition of misconduct included a second clause referring

to "material failure to comply with federal requirements that uniquely relate to the conduct of research." This clause was eliminated in the misconduct definition adopted in the final rule (DHHS, 1989a) to avoid duplicate reporting of violations of research regulations involving animal and human subjects, since these areas are covered by existing regulations and policies.

9. In the commentary accompanying its final rule, NSF (1987) noted that several letters on the proposed rule had commented that the proposed definition was too vague or overreaching. The NSF's 1987 definition originally included two clauses in addition to those in the PHS misconduct definition: "material failure to comply with federal requirements for protection of researchers, human subjects, or the public or for ensuring the welfare of laboratory animals" and "failure to meet other material legal requirements governing research" (NSF, 1987, p. 24468). These categories were removed in 1991 when the regulations were amended.

10. In a "Dear Colleague Letter on Misconduct" issued on August 16, 1991, the NSF's OIG stated, "The definition is not intended to elevate ordinary disputes in research to the level of misconduct and does not contemplate that NSF will act as an arbitrator of mere personality clashes or technical disputes between researchers."

11. K. Louis, J. Swazey, and M. Anderson, *University Policies and Ethical Issues in Research and Graduate Education: Results of a Survey of Graduate School Deans*, preliminary report (Bar Harbor, Me.: Acadia Institute, November 1988). The survey was published as Swazey et al. (1989).

12. It should be noted that the survey instrument used by the Acadia Institute did not define "research misconduct," but instead left that term open to the interpretation of the respondents. In some parts of the survey, "plagiarism" was distinguished from "research misconduct."

13. Sigma Xi (1989), as summarized in NSF (1990d), pp. 4-5.

14. Cited in Woolf (1988a), p. 71. She quotes an editorial by Koshland (1987) for the first figure and a survey by St. James-Roberts (1976b) for the latter.

15. See Tangney (1987) and Davis (1989). See also St. James-Roberts (1976a). The reader survey reported in St. James-Roberts (1976b) received 204 questionnaire replies. Ninety-two percent of the respondents reported direct or indirect experience with "intentional bias" in research findings. The source of knowledge of bias was primarily from direct contact (52 percent). Forty percent reported secondary sources (information from colleagues, scientific grapevine, media) as the basis for their knowledge.

See also *Industrial Chemist* (1987a,b). The editors expressed surprise at the high level of responses: 28.4 percent of the 290 respondents indicated that they faked a research result often or occasionally.

5

Handling Allegations of Misconduct in Science—Institutional Responses and Experience

UNIVERSITY–GOVERNMENT APPROACHES

Growing Interaction in the 1980s

Public discussions of cases involving misconduct in science are common today in research seminars or professional meetings, but such discussions were rare until the past decade.[1] Congressional hearings convened in 1980 by the House Science and Technology Committee's Subcommittee on Oversight and Investigations, chaired by then Rep. Albert Gore, were the first systematic public examination of reports of fraud in biomedical research (U.S. Congress, 1981a).

Prior to the mid-1980s, academic institutions sometimes examined allegations of misconduct in science through faculty conduct committees or other disciplinary procedures. But reports of formal investigations were rarely communicated to research sponsors, editors, or other research scientists; more commonly, misconduct-in-science cases were handled privately—if they were handled at all. Many universities adopted procedural reforms for addressing misconduct in science after a series of highly publicized cases in the early 1980s revealed the shortcomings of institutional processes for dealing with cases involving federal research funds (U.S. Congress, 1981b).

In 1985 Congress passed legislation (P.L. 100-504) requiring institutions that receive Public Health Service (PHS) research funds to adopt an administrative process for reviewing reports of scientific fraud. This legislation also authorized the secretary of the Department of Health and Human Services (DHHS) to adopt regulations to require assurances from PHS grantees that such an administrative process was in place (DHHS, 1986). In fulfilling its obligation under the legislation, the PHS issued interim guidelines in 1986 on policies and procedures for handling alleged misconduct. Educational and scientific societies such as the Association of American Universities, the Association of American Medical Colleges, and the American Association for the Advancement of Science provided forums to review progress by the research institutions in complying with the new regulatory requirements. In addition, educational, legal, and scientific organizations suggested approaches to addressing allegations of misconduct in science, assessing institutional experience in handling misconduct complaints, and providing national and regional forums for the exchange of information and experience in this area (AAU, 1983; AAMC, 1982; AAAS-ABA, 1989).

In 1989 the DHHS's Office of Inspector General (OIG) reported that most of the research-intensive universities (institutions with 100 or more PHS awards) had adopted formal policies and procedures for addressing allegations of misconduct in science but that only 22 percent of all PHS institutional grantees had such policies and procedures (DHHS, 1989d). The OIG report suggested that many of the institutional policies and procedures were limited in scope and that most did not require notification to the National Institutes of Health (NIH) as the 1986 interim guidelines had suggested (DHHS, 1989d).[2] The report found that some grantee institutions had been waiting for final PHS regulations to be promulgated before developing their own policies and procedures.

Following congressional hearings that criticized university and agency responses to allegations of misconduct, the PHS, in 1989, published final regulations requiring all applicant and grantee institutions to adopt misconduct policies and procedures (42 C.F.R. 50).[3] Although the regulations provide general direction about procedural elements (such as requirements for an inquiry, an investigation, disclosure, and notification), the specific content and scope of misconduct policies and procedures remain at the discretion of the research institutions. The procedural elements have been the subject of extensive debate and discussion over the past 3 years as more experience has been acquired by research institutions and federal agencies in the implementation of the regulations.

General Requirements

All research institutions that receive PHS funds must now provide assurances that they have adopted policies and procedures to handle allegations of misconduct in science. NSF also requires that a grantee institution have such policies and procedures if that institution wishes NSF to defer to it for purposes of inquiry and investigation of misconduct cases. Because research institutions are able to design their own misconduct policies and procedures, institutional responses to federal regulatory requirements are very diverse. At present consensus is lacking about which procedural approaches are adequate responses to federal regulatory requirements, and institutional and governmental officials frequently disagree over fundamental matters of openness, completeness, or timeliness.[4]

Institutions that receive PHS research awards are required to submit to the DHHS's Office of Scientific Integrity Review (OSIR) an initial assurance and annual reports of compliance indicating that they have adopted policies and procedures for handling allegations of misconduct in science. PHS officials review research and training grant applications to determine whether the institutional assurance requirement has been met and may request copies of the institution's policies for addressing misconduct in science. However, they do not certify the acceptability of such institutional policies. PHS officials have judged some institutional investigative reports to be inadequate, even though the reports complied with local institutional policy and procedures for handling misconduct in science.

More Specific Requirements Related to Misconduct Policies and Procedures

Government regulations require that institutional policies and procedures include two separate stages: an inquiry and an investigation. An inquiry—a preliminary review of the complaint and other information to determine if there is sufficient basis for an investigation of alleged misconduct—does not yield a judgment on the question of guilt, although it can determine that an allegation lacks merit.[5] An investigation is a formal examination and evaluation of relevant information to determine whether misconduct has occurred. Such an investigation, often using a standing committee or an ad hoc panel of experts, produces a report that includes findings, and possibly recommendations, that form the basis for an adjudicatory decision by a responsible institutional official. In cases where institutions find misconduct in science, government officials may recommend penalties including institutional oversight, certification of future research applications, prohibition from service on government committees, or debarment. Government agencies may initiate separate proceedings to adjudicate cases involving serious offenses for which severe sanctions, such as a recommendation for debarment, are to be considered.

Research institutions must inform the sponsoring agency in writing when the institution decides to move from an inquiry to an investigation of an allegation of misconduct in science (DHHS, 1989a). A final report of the misconduct investigation must be submitted to the sponsoring agency, and the report may be subject to disclosure in response to requests under the Freedom of Information Act.

Both NSF and PHS rely on research institutions to conduct misconduct inquiries and investigations. But if government officials determine that the report of an institutional inquiry or investigation is not thorough, fair, objective, or responsive to government regulatory requirements, the agencies may intervene and investigate allegations of misconduct directly. The criteria for determining what constitutes an "adequate" inquiry or investigation remain somewhat vague, although PHS and NSF officials have made efforts to clarify the policies, procedures, and criteria that guide their evaluations of investigative reports.[6] [7]

Institutional Responses to Requirements

Before government agencies adopted regulations for handling allegations of misconduct in science, most universities and other research centers addressed such complaints through a variety of informal and formal, often confidential processes. In the early 1980s, few academic institutions had formal policies or procedures to review allegations of misconduct in science. For example, a 1982 survey indicated that fewer than one-quarter of the respondent academic institutions and hospitals had written rules to deal with allegations of fraud but that just over one-half were reportedly engaged in formulating such rules.[8] The survey found "vast differences of opinion" (p. 214) about the desirability or necessity of rules or policies for responding to allegations of fraud in research as well as "major disagreements" (p. 207) about the issues to be addressed by such policies (Greene et al., 1985). Since that time, many academic institutions have adopted policies and procedures for handling allegations of misconduct in science, but substantial variation remains in definitions and methods for conducting inquiries and investigations.

Although the DHHS's OSIR and the NSF's OIG have evaluated reports of some misconduct investigations,[9] the experience of research universities in conducting misconduct-in-science inquiries and investigations has not been comprehensively analyzed. Thus information about the broad range of experience of diverse institutions in handling allegations of misconduct in science is often derived from anecdotes and journalistic accounts describing the experiences of universities and individual participants in specific cases.

Stage One: Misconduct Inquiry

Research institutions have different methods for structuring an inquiry to determine whether an allegation of misconduct in science has substance.[10]   Some institutions (such as Harvard Medical School) rely on existing faculty conduct committees to handle misconduct inquiries and investigations, if necessary.   Others (such as The University of Chicago) have established a standing committee on academic fraud to oversee the university's handling of misconduct cases. Some universities (such as the University of California, San Diego) rely on administrative officials to appoint an investigator or faculty panel to conduct a preliminary inquiry and subsequent investigation, if necessary.

When an initial allegation of misconduct has been made, an administrative official or designated faculty member usually conducts a confidential inquiry in response to the allegation. The official may consult with selected faculty members or co-workers to determine the nature of the suspected offense, and, in some cases, the individual accused of misconduct may not be informed that an allegation has been made. The inquiry may be closed by the preparation of a brief file memorandum—which may be provided to the complainant for comment—that either states the reasons that no further investigation was judged to be necessary or recommends that an investigation be initiated.

Stage Two: Misconduct Investigation

If an investigation is recommended, the individual accused of misconduct and the appropriate government research sponsor are informed of the nature of the allegations. According to an OSIR analysis, research institutions generally establish a panel of scientific experts, usually numbering three to eight members, to conduct an investigation, review evidence, and interview witnesses and relevant parties (DHHS, 1991b). For the most part, such a panel is composed of persons from the institution, although they are usually not from the department or research center of the subject of the investigation. In a few cases, institutions have used panel members from other organizations. The OSIR's report describes the investigative process as follows (DHHS, 1991b, p. 4):

> The typical investigation conducted by institutions involved interviews of the subject, the informant, and other relevant parties, review of publications, manuscripts, or other documents, review of data notebooks, and in a few cases, site visits to the laboratories involved. In 5 of the outside [i.e., non-PHS] investigations, the subject was accompanied by legal counsel during meetings with the panel, with counsel acting in an advisory capacity. Interviews were recorded or transcribed in 6 of the outside investigations and in 2 of the PHS investigations.
>
> One institution held a formal hearing before a five-member "Hearing Board." Provision was made for full disclosure of evidence prior to the hearing, testimony from witnesses, cross examination of witnesses by the subject's attorney, and written and oral summary positions at the end of the hearing. The hearing transcript was made available within ten days and the record was kept open for about two weeks to allow for additional information or comment.
>
> The time required for outside institutions to complete investigations varied from one to 12 months. The majority of the outside investigations for which PHS accepted the conclusions were completed within 4 months (10 out of 16). One of the investigations conducted by PHS components took 12 months, one lasted 9 months, one 7 months, one 3 months, and one 2.5 months.
>
> There does not appear to be any systematic relationship between the nature of the alleged misconduct and the amount of time required to conduct an investigation.

In misconduct cases reviewed by PHS and NSF, research institutions have sometimes imposed sanctions as a direct result of their investigations, in some cases prior to or in addition to governmental actions. Private settlements between a research institution and an individual accused of misconduct have also been reported, in which an individual accused of misconduct agrees to resign in lieu of the institution initiating a formal investigation, although such settlements may not be consistent with government regulations. Cases involving serious offenses that could result in dismissal or termination of funding usually have clear distinctions between the investigative and adjudicatory stages. Most academic institutions have specific procedures that must be followed before severe disciplinary sanctions can be applied to tenured faculty. These procedures typically are invoked after the misconduct investigation concludes that serious misconduct did occur and that disciplinary action is warranted. In some cases, the investigative panel may then refer the case to a faculty conduct committee or an academic senate for adjudication. But am-

TABLE 5.1 Types of Local Institutional Actions Resulting from Misconduct Investigations, March 1989 to December 1990

Penalty or Action	Number of Cases
Issued letter of reprimand	1
Terminated research support (i.e., would not allow subject to continue as principal investigator)	2
Required review of future applications for research support	1
Informed future prospective employers of findings	1
Required correction of literature or withdrawal of manuscripts	3
Denied or revoked tenure	2
Dismissed subject or requested retirement	4[a]
Accepted voluntary retirement	1

[a]Includes dismissal of an NIH intramural scientist.

SOURCE: Department of Health and Human Services (1991b).

biguity remains as to whether an individual accused of misconduct in science is entitled to a full disciplinary hearing for penalties or disciplinary sanctions that may be mild, such as a letter of reprimand or mandatory supervision.

The types of institutional actions taken in response to misconduct investigations reviewed by the DHHS's OSIR are given in Table 5.1.

Findings, Discussion, and Conclusions

Government agencies, congressional oversight committees, and academic institutions generally agree that *the primary responsibility for handling complaints of misconduct in science rests with the research organization.* However, the development and implementation of policies and procedures for handling misconduct in science have been problematic. Some universities, particularly small research institutions, are not prepared to accept responsibility for pursuing allegations of misconduct in science.[11] It is difficult for any institution to investigate members of its own community, especially individuals who hold positions of high esteem. In addition, some research institutions and government agencies have made mistakes in investigations of complex cases, such as appointing to investigatory panels members who have personal or professional ties to the individuals who have been accused of misconduct in science. All these factors foster a perception that research institutions are not dealing effectively with misconduct in science,[12] prompting criticism of the speed, rigor, honesty, fairness, and openness of the mechanisms now used by academic institutions to address misconduct in science. Public officials, journalists, and even some scientists themselves continue to question whether universities are willing to address the problem of misconduct in a vigorous and effective manner.[13]

However, the difficulties of maintaining informed awareness of existing policies and procedures in the academic research environment should not be underestimated. Scientists and students are highly mobile, and research centers are decentralized organizational units within the university. Informing individuals about appropriate methods for raising concerns about misconduct in the research environment requires sustained collaboration among research administrators, faculty, and laboratory directors.

Many universities have now established policies and procedures for handling allegations of misconduct in science, and some research institutions have acquired valuable experience in implementing these procedures to deal with cases of misconduct. However, the legal and procedural issues associated with misconduct-in-science investigations are extraordinarily complex, and there is little case law in the public record to guide and inform analysis of these issues.

The panel believes that, in general, the current and evolving system of government and institutional relationships requires more experience and adjustments before specific policy or procedural changes can be recommended. Research institutions and government agencies need to clarify their own approaches and judgments on these issues before any general consensus can be reached on procedural matters. The panel did not have a sufficient base of institutional experience or consensus about these matters within the academic community on which to develop recommendations about the nature of institutional procedures for handling allegations of misconduct in science.

Part of the difficulty in developing vigorous and effective institutional responses to incidents or allegations of misconduct in science arises from variation in and disagreement about essential elements of fairness, completeness, and objectivity that should characterize investigations. Effective responses are impeded also by recurring patterns of denial by some institutional officials and faculty members who believe that misconduct in science is not a serious matter. The pressures of conducting an objective investigation of complaints involving respected or prestigious scientists cannot be underestimated.

Strong and informed leadership is needed to clarify procedural matters and to ensure that allegations or apparent incidents of misconduct in science are not ignored or covered up.

Members of the research community and government officials agree that deliberate efforts to misrepresent research findings or to distort the research process should not be tolerated. Disagreement focuses on which acts of misconduct should be subject to institutional or governmental penalties and what methods are appropriate to respond to unprofessional behaviors that do not fit institutional or regulatory definitions of misconduct in science.

Experience suggests that complainants, administrative officials, or investigative panels may be unable to determine, at the outset, whether the behavior in question constitutes misconduct in science, other misconduct, a questionable research practice, or none of these. Allegations of misconduct are sometimes based on uncertain or fragmentary information, and the nature of the suspected offense may change as additional evidence is obtained.

Whatever procedures are adopted, the point of first contact and early judgment in handling allegations of misconduct in science are extremely important. Although it is necessary to preserve informality and flexibility in handling individual complaints, some of which may be unfounded or mistaken, it is also important to assure the credibility of the process by which these complaints are addressed.

General Conclusions

Institutional policies and procedures should include a common entry point for handling complaints from the outset; clear procedures are necessary for determining which types of alleged offenses will be reviewed by administrative staff or faculty. A sequence of steps to achieve resolution of significant disputes is required. All of these steps require clear separations between each of the following groups: the affected parties, those who are judging the seriousness of the complaint and formulating the evidentiary base to substantiate charges, and those who must adjudicate penalties based on charges of misconduct in science.

Disputes or accusations involving questions of scientific judgment or questionable research practices are generally settled, whenever possible, by the research process itself. However, when disputes involve specific charges of misconduct in science or other misconduct, they cannot be resolved by scientists alone. Institutional procedures, based on sound legal principles, are necessary to determine whether such an allegation has substance and, if so, to implement appropriate responses and penalties, if warranted.

The appropriate treatment of misconduct allegations is time consuming and costly, and it diverts faculty and administrative attention from other matters. Questions about the integrity of an individual also create enormous emotional stress; at least two incidents of suicide have been associated with the investigation of misconduct allegations.

The Issue of Adjudication

Some authors have noted that confusion exists about the nature of the investigative stage in both university and governmental investigations of misconduct in science.[14] Criminal and civil legal procedures traditionally distinguish between "investigations" and "adjudications" for purposes of due process analysis (Andersen, 1988). "Investigations" are commonly thought to be fact-gathering processes that precede formal charges. "Adjudications" are deliberations as to the guilt or innocence of the individual who has been charged. However, many institutional policies and procedures for addressing misconduct in science do not specify this distinction. Thus in some cases, findings of guilt or innocence, rather than charges of misconduct, may result from an investigative panel's deliberations, leading to criticism that appropriate due process concerns were not met in the investigation.

As a result, the amount of confidentiality appropriate for the investigative stage has not been clearly resolved. Research institutions are required by NSF and PHS regulations to inform the research sponsor when investigations have been initiated, and some observers have suggested that moving from an inquiry to an investigation is thus comparable to an indictment by the courts. Many individuals in the scientific community have complained that the reputation of a subject of a misconduct investigation is damaged simply by the announcement that an investigation has been initiated—before the completion of the investigation and before the subject of the investigation has had an opportunity to confront witnesses or respond to evidence. Confusion is also compounded by the fact that many scientists and others view the imposition of formal charges of misconduct in science as a de facto adjudicatory decision.

The panel believes that institutional procedures should define explicit and clear criteria that are to be used in determining when a misconduct inquiry should proceed to a more formal investigation. The panel concludes that administrative officials and faculty have a responsibility to inform all members of their institution, especially junior personnel, of existing channels for handling complaints about misconduct in science or other misconduct.

GOVERNMENT REGULATIONS AND PROCEDURES

The Public Health Service and the National Science Foundation have promulgated procedures for the federal agencies themselves in addressing charges of misconduct in science. The federal procedures govern the handling of such charges by the agencies and also serve as a model for universities. Federal procedures are invoked if the university has not investigated alleged misconduct or if the funding agency concludes that the university investigation was inadequate. Federal procedures may also be invoked if the grantee institution lacks the resources and the impartial personnel needed to conduct an investigation. These procedures also apply if the funding agency seeks to impose additional sanctions (Andersen, 1988).

During the period of the panel's study, various administrative and legislative proposals were introduced to organize the government's activities in handling allegations of misconduct in federally supported research programs. The panel was not able to review fully each of these proposals, since some were published late in the deliberative stages of the study.[15] Recognizing the evolving character of the organizational programs designed to address misconduct in science, especially in the PHS, the panel did not attempt to define specific procedures for federal agencies or the relationship of individual offices but focused instead on issues pertinent to the roles and responsibilities of government in handling allegations of misconduct in science.

The Health Research Extension Act of 1985 (P.L. 100-504) established legislative authority for PHS regulations and other policies for identifying incidents of misconduct in science involving the use of federal funds. Final regulations were promulgated by PHS in 1989 (DHHS, 1989a). In the intervening years, there was much confusion and uncertainty about the nature of the required policies, the definitions of misconduct that should be incorporated into these policies, and the relationship of institutional responsibilities to those of the oversight agencies. NSF regulations, adopted in July 1987 and revised in May 1991 (NSF, 1987, 1991b), differ from the PHS regulations in some significant matters.

Department of Health and Human Services

The responsibility for handling misconduct in science is divided between two offices in the Department of Health and Human Services (DHHS)—the Office of Scientific Integrity (OSI) and the Office of Scientific Integrity Review (OSIR). The DHHS's Office of Inspector

General (OIG) has prepared studies relevant to issues of misconduct in science and may also open an investigation of cases that may involve criminal behavior. In addition, the PHS has appointed misconduct policy officers in each funding component.

Office of Scientific Integrity

The Office of Scientific Integrity, created in May 1989, is the administrative unit that has oversight responsibilities for implementing the PHS policies and procedures related to misconduct in science (DHHS, 1990a). Located in the office of the NIH's director, OSI reviews misconduct-in-science allegations to determine whether sufficient information exists for an institution (or OSI, if the allegation cannot be referred to an institution) to conduct an inquiry under PHS regulations. OSI monitors investigations conducted by institutions that receive PHS funds for biomedical or behavioral research.

The OSI's policy is to conduct its own inquiry or investigations "if an institution has demonstrated an inability or unwillingness to conduct a thorough and objective inquiry or investigation or if the institution's inquiry or investigation does not adequately resolve the issue" (DHHS, 1990a, p. 8). OSI also carries out misconduct investigations if PHS intramural research personnel are the subjects of a complaint that has been substantiated after an inquiry.

The OSI provides the subjects of misconduct investigations with an opportunity to review and comment on the investigative report and findings, as well as sanctions that may be proposed, before OSI sends its findings to OSIR. All these comments become a part of the record considered by the OSIR in its review of the case.

Office of Scientific Integrity Review

The Office of Scientific Integrity Review is a component of the office of the assistant secretary for health, who also serves as the head of the PHS. OSIR establishes overall PHS policies and procedures for addressing misconduct in science and reviews final reports of misconduct investigations (both governmental and institutional) to ensure objectivity and fairness. When misconduct in science has been established by OSI, OSIR makes final recommendations to the assistant secretary for health regarding any sanctions to be imposed by PHS. If debarment is recommended, the assistant secretary will forward this recommendation to the DHHS's debarment official, who provides an opportunity for a formal hearing by the subject of the proposed debarment.

Office of Inspector General

The DHHS's Office of Inspector General has responsibilities for investigating complaints about waste, fraud, and abuse involving DHHS funds in areas such as Medicare and Medicaid payments and student loans. In a 1988 report on the handling of allegations of misconduct in science, OIG criticized the arrangements and procedures used at that time by the PHS and recommended that responsibilities for these matters be centralized (DHHS, 1989d). This report preceded the formation of OSI and OSIR.

The OIG provides specialized expertise and authority to OSI and OSIR in their efforts to address misconduct in science. If criminal behavior is suspected, OIG may issue subpoenas or provide access to restricted information in investigating charges of misconduct in science, and it has done so in at least one case (DHHS, 1991b).

PHS ALERT System

The Public Health Service currently maintains the PHS ALERT system, which is a system of records identifying individual investigators and institutions that are under investigation for possible misconduct in science or who are subject to penalties for such misconduct. As of mid-January 1991, the PHS ALERT system had confidential records for 81 individuals and 5 institutions. Responsibility for maintaining and managing the PHS ALERT system rests with OSI. OSI searches the PHS ALERT system on a regular basis to compare the records it contains with the list of PHS grant recipients. The name of an investigator on file in the PHS ALERT system may be submitted to the funding directors of an institute, who may use the information in making decisions about, for example, advisory committee appointments and grant extensions.

Conflicting Views About Use of the PHS ALERT System. The identification and possible mistreatment of individuals who are subjects of ongoing but unresolved investigations have been criticized by many scientists. The notification provided by the PHS ALERT system can jeopardize the award of PHS research funds and government advisory appointments. Reputations can be damaged by use of the PHS ALERT system prior to a determination of misconduct, and some misconduct investigations can take several years to complete. But government officials note that access to the PHS ALERT system is restricted and contend that agency directors should have the opportunity to be informed that a misconduct investigation is in process prior to awarding research funds or making advisory committee appointments involving a subject of such an investigation.

Panel Findings and Conclusions About Use of the PHS ALERT System. The conflicting views about issues related to confidentiality were considered by the panel. Fairness requires that the subject of misconduct investigations should have an opportunity to respond to charges and evidence before the findings of the investigation are communicated to others. In some cases, the first public notice of a misconduct-in-science affair has come with the release of a draft report of an investigation, before the subject has had an opportunity to respond. This situation cannot be tolerated.

The use of the PHS ALERT system in disclosing the identities of individuals who are under investigation for possible misconduct in science is a serious flaw in the fairness of current governmental policies and procedures. It is possible that incomplete information and unsubstantiated allegations may jeopardize research awards or governmental appointments and that individual scientists may be victimized by premature release of draft investigative reports. The panel concludes that government agencies should suspend the practice of disseminating notices of misconduct-in-science investigations in the PHS ALERT system until formal charges of misconduct of science have been filed.

National Science Foundation

Responsibility for handling investigations and monitoring allegations of misconduct in science in NSF programs and operations rests with the NSF's OIG. This office, established by the Inspector General Act Amendments of 1988,[16] also has responsibility for handling audits of grants, contracts, and cooperative agreements funded by NSF, the financial misconduct of employees in connection with their duties, as well as conflicts of interest involving NSF programs. Responsibility for adjudication of findings of investigatory reports resides with the Office of the NSF Director. Working with the general counsel and the National Science Board, the NSF's director formulates NSF regulations on misconduct in science, often in coordination with the Office of Science and Technology Policy and the PHS. The OIG implements the portion of the regulations that have to do with investigating misconduct in science. It publishes a semiannual report each year for the Congress documenting its efforts and providing summary data as well as specific examples of misconduct-in-science cases that have been addressed by the office.

The NSF's policy is that research institutions should be responsible "to the greatest extent possible" for preventing and detecting misconduct in science and for dealing with any allegations of misconduct that may arise (NSF, 1991a, p. 30). The NSF expects research institutions to conduct inquiries and investigations, if warranted, into incidents of suspected or alleged misconduct. The NSF's policy uses the concept of "deferral in the first instance" in establishing its relationships with the research community. This policy recognizes both the institution's commitment to maintain integrity in research and the independence and autonomy society accords the research community. However, it also places a critical obligation on an institution that requests and accepts deferral. The institution is obliged to conduct an investigation that OIG can recognize as accurate and complete. OIG must also be able to conclude that fair and reasonable procedures in accord with due process were followed (NSF, 1991a, p. 31). NSF regulations, which share general similarities with but differ from PHS regulations for addressing allegations of misconduct in science, establish procedural requirements but rely on research institutions to establish their own policies and procedures.

GOVERNMENT–UNIVERSITY EFFORTS— UNRESOLVED ISSUES

The role of government agencies in handling alleged or suspected misconduct in science has been the subject of extensive examination within the academic and research community, government agencies, and the Congress. Although there is strong consensus favoring the principle that universities should bear the primary responsibility for addressing misconduct in science, there is substantive disagreement about the methods by which this responsibility should be exercised and the manner in which federal agencies should perform oversight.

Areas of Disagreement

The areas of disagreement include the following:

• *Definitions of misconduct in science.* Government regulations offer general definitions of misconduct in science but do not provide extensive guidance about the scope of the definitions (e.g., defining fabrication, falsification, or plagiarism). Institutional officers, faculty, and public officials sometimes disagree about specific behaviors that constitute misconduct in science. Disagreement over definitions of misconduct in science in governmental and institutional policies and procedures can lead to uncertainty about whether to include as misconduct in science those cases that involve charges of incompetence, science conducted with "reckless disregard" for the truth, and other misconduct, such as sexual harassment, that may occur in the research environment.[17]

• *Nature of evidentiary findings.* University and government officials sometimes differ on the nature of evidence that is necessary to substantiate an allegation or suspicion of misconduct in science. For example, some institutions have concluded that carelessness and poor judgment do not constitute misconduct in science. Government officials have sometimes disagreed with such findings, particularly when, in the government's view, there was evidence to show that deception was intentional. In other cases, government officials have criticized or rejected institutional reports of inquiries or investigations as "defective" when these reports lacked sufficient information to enable others to assess the fairness or completeness of the institutional process.

There are also differences in evidentiary standards used by research institutions and government agencies. Institutional officials may prefer to use a higher standard of "clear and convincing" evidence or evidence that is "beyond a reasonable doubt," while government officials may rely on a less restrictive "preponderance-of-the-evidence" standard to substantiate a finding of misconduct in science.

• *Due process requirements.* There is confusion about the formal procedures that are required in the resolution of allegations of misconduct in science. Since government officials often rely on institutional investigative reports in recommending possible sanctions, there can be different expectations and standards of procedural clarity, fairness, and objectivity. The OSI's approach, which has been criticized (Hamilton, 1991a), maintains what it calls the "scientific dialogue" model of investigation, in contrast to what OSI staff term the "legal-adversarial" approach adopted by NSF, in the belief that the former method can both reveal the scientific facts of the case and also secure the due process rights of the respondent without inviting the difficulties of adversarial proceedings.[18]

Some of the policies and procedures used by OSI in its investigations and oversight have been challenged in the courts and criticized in the press. Several problems have been identified: the inability of the subjects and key witnesses of the investigations to review all evidence until the compilation of the investigative report, premature disclosure of draft reports in the press, and the absence of disclosure of such draft reports to institutional officials who may be affected by the outcome of the investigation.

• *Quality and timeliness of investigations.* The NSF and PHS have the authority to conduct their own investigations of alleged misconduct in science, if the institutional reports are judged to be inadequate. Some academic institutions believe that government agencies have been overly intrusive or have disrupted academic investigations, especially when public officials have intervened before a university has completed its investigation. Both university and government investigations have been criticized because of the lengthy period required to complete inquiries or investigations in some misconduct cases.

• *Leaks of draft reports.* In March 1991, draft reports of two misconduct-in-science cases under investigation by OSI were leaked to the press.[19] According to OSI policy at that time, confidential drafts of the reports were circulated to principals in the case, including subjects, complainants, and institutional officials. Draft reports of misconduct investigations often provide the first opportunity for subjects to review statements and other evidence used in evaluating allegations against them. The draft reports may be incomplete because they lack additional information that can be provided by the subject or others in responding to such statements.

Leaks of draft investigative reports represent a serious breach of confidentiality and procedure that could prejudice not only the outcome of particular cases, but also the fairness and security of OSI procedures. OSI has taken steps to reduce the possibility of leaks by a change of procedures, whereby significant witnesses (such as the original complainant) have only limited and supervised access to the draft investigative report.[20] However, the possibility remains that individuals accused of misconduct may leak draft investigative reports to serve their own interests. The damage to reputation that may occur from public disclosure of draft investigative reports imposes greater requirements for assurances that subjects accused of misconduct will have opportunities to respond to charges and testimony prior to the preparation of the draft report.

Observations and Discussion

All of the issues listed above require attention. The quality of institutional investigations of allegations of misconduct in science might be enhanced by a critical examination of procedural experiences—especially the opportunities to respond to charges, question witnesses,[21] and comment on draft reports—derived from handling forms of other misconduct in the academic environment. For example, information on the disciplinary procedures used to address complaints of incompetence or alleged violations of academic codes of conduct by faculty members or students might be helpful in understanding how to deal with misconduct in science. Lessons could also be derived from the experiences of other federal agencies in the investigation of charges of Medicare or Medicaid fraud; charges alleging fabrication or falsification of scientific data supporting new pharmaceutical products and devices examined by the Food and Drug Administration; and alleged violations of contractor performance in space, defense, agriculture, or energy-related research programs. In exceptional cases, when deliberate institutional cover-ups of misconduct in science investigations are suspected, governmental responses should be guided by the same practices that govern cover-ups of contractor fraud or financial misconduct.

The success of interactions among scientists, university representatives, and government officials in handling allegations of misconduct in science can be assured only if all groups agree about actions that constitute misconduct in science and make a commitment to addressing misconduct in science by invoking consistent, firm, and fair procedures.

THE ROLE OF THE COURTS

The courts have become centrally involved in disputes arising from allegations and investigations of misconduct in science. A researcher who is the subject of an allegation of misconduct may seek judicial examination of the actions of the government or the university following or prior to the completion of an investigation of the allegations. Those whose interests may be affected by misconduct—for example, those with interests in intellectual property that has been appropriated or those with copyright interests—may bring the subject of the allegation before the courts. And in particularly egregious cases of misconduct, the government may pursue criminal charges against the researcher.[22]

Only in a few cases have the courts imposed criminal sanctions on scientists found guilty of misconduct in science and other violations of research regulations. The courts have imposed financial penalties as well as requirements for community service.[23] More recently, a court has considered policies and procedures used by the PHS to guide daily operating processes in addressing allegations of misconduct. The court concluded that these statements had not been validly promulgated, and the case is now on appeal.[24]

Complainants in a misconduct dispute also have the right to involve the courts. Federal law provides a cause of action termed a *"qui tam* action" in which a private citizen may bring an action on behalf of the United States to recover government funds. The private individual may be allowed in such a case to receive a portion of those funds as a reward for pursuing the litigation. Such actions have arisen in the context of misconduct in science cases, and the courts have become involved in reviewing *qui tam* claims on several occasions (Cordes, 1990).

SPECIAL CONCERNS PROMPTED BY UNIVERSITY–GOVERNMENT–COURT INTERACTIONS

Five issues require special consideration in examining interactions among research institutions, government agencies, and the courts in the handling of allegations of misconduct in science:

1. Due process requirements for fair and objective institutional investigations of alleged or suspected misconduct in science,
2. The consequences of misconduct inquiries and investigations,
3. Faculty participation in misconduct investigations,
4. The role of whistle-blowers, and
5. The problem of false allegations in misconduct investigations.

Due Process Requirements

The due process clause of the Fifth and Fourteenth Amendments of the U.S. Constitution requires that the government follow fair procedures before depriving an individual of "life, liberty or property."[25] The purpose of procedural due process is not only to "prevent unfair and mistaken deprivations" of constitutionally protected interests,[26] but also to allow affected persons to participate in a decision of vital importance to them.[27] If an affected interest is at stake, the Constitution requires that the decision must be made using fair procedures.

The due process clause applies only to "state action." Thus the constitutional limitations directly affect decision making only by governmental entities—in this case, the funding agencies or state universities. Private universities may have constraints on their decision-making processes that arise from contractual relationships with faculty and staff that are similar to those imposed by the Constitution. Hence the requirements of due process provide the benchmark against which misconduct procedures should be evaluated.

An individual may have a property interest in a research or faculty position, particularly when the expectation of continued employment is explicitly granted by tenure or contract.[28] Moreover, a government debarment action or the suspension of research funds might be seen as deprivations of a property or liberty interest. If so, government investigations must follow stringent procedures to minimize erroneous findings and meet the requirements of due process. However, less severe penalties, such as a letter of reprimand or a requirement of prior approval for particular activities, are probably not deprivations of constitutionally protected interests. While these sanctions might injure a scientist's reputation, such injury, absent a change in job status, is not recognized by the Supreme Court as a deprivation of a constitutionally protected interest.[29]

The Supreme Court has developed a balancing test to determine specific procedures that must be employed before an individual may be deprived of a constitutionally protected property or liberty interest.[30] On the side of the accused the Court weighs the importance of the liberty or property interest at stake and the extent to which the procedure at issue may reduce the possibility of erroneous decision making. On the other side, the Court considers the government's interest in not increasing its administrative and fiscal burdens.

Constitutionally required procedures are defined by a balancing process, and detailed requirements emerge through case decisions. Although constitutional protections apply only to actions by the government (i.e., a funding agency or a state university), the need for a fair process applies to any resolution of a case alleging misconduct in science. In order to accord with the principles of fairness embodied in due process, procedures for resolving misconduct-in-science cases probably should contain the following elements:[31]

1. A clear specification of what constitutes misconduct, as well as the possible sanctions.

2. Assurance that when misconduct in science is alleged or suspected, an initial inquiry will be made to determine if a hearing[32] is warranted. This inquiry should remain confidential in order to protect the reputation of the accused from groundless or trivial charges. It is not necessary to notify the accused or the research sponsor of the inquiry.

3. Stipulation that if the evidence gathered from the initial inquiry warrants a hearing, notice will be given to the accused and the research sponsor of the charge and of the conduct or transaction(s) on which it is based, as soon as possible, consistent with the protection of evidence, particularly in potential criminal cases.

4. Provision of a hearing conducted by impartial decision makers.

5. Prompt completion of the initial inquiry and hearing.

6. Assurance that, at a minimum, an individual found guilty of misconduct in science will be provided with the investigation report and given an opportunity to file a written objection—procedures that would generally be sufficient for mild sanctions such as a reprimand (without a change in job status), special monitoring of future work, or probation.

7. In hearings that consider more severe sanctions (suspension, salary reduction, rank reduction, or termination of employment), a requirement for many if not all of the following additional procedures on behalf of the subject of the misconduct allegations: (a) the opportunity to make an oral presentation to the decision maker; (b) the opportunity to present evidence or witnesses to the decision maker; (c) the opportunity to confront the witnesses against the accused and/ or to review the documentation that serves as evidence of the allegations; (d) the right to have an adviser to assist in presenting the accused's case to the decision maker;[33] and (e) a decision based on the record with a statement of reasons.

8. A statement of exoneration should be issued if misconduct is not established.

Consequences of Misconduct Inquiries and Investigations

An investigation can result in a finding of "misconduct" or "no misconduct." Research institutions, government and other sponsors, editors, prospective employers, and others may take actions as a result of a finding of misconduct. But if an inquiry or investigation does not establish a finding of misconduct in science or identifies problems that do not meet the criteria for a finding of scientific misconduct, the research sponsor or research institution may still take remedial actions.

Government Sanctions

The results of a misconduct investigation must be reported to the government research sponsor, and the sponsor may then determine what, if any, sanctions should be imposed. In determining the appropriate sanction for a particular act of scientific misconduct, government agencies consider (1) the seriousness of the misconduct, (2) whether it was a deliberate or merely a careless action, (3) whether it was an isolated event or part of a pattern, and (4) whether it is relevant only to certain funding requests or awards or to all requests and awards of the accused.[34]   The burden of proof is on the agency proposing the sanctions, and the agency must prove its case by a preponderance of the evidence.[35]

The NSF groups its possible sanctions into three classes, ranging from the least restrictive (such as a letter of reprimand) to the most severe (including termination of a grant and recommendation for debarment).[36] Individuals subject to less severe restrictions are entitled to fewer procedural safeguards, whereas procedures for imposing debarment are strictly defined.

The PHS categories for sanctions for misconduct differ slightly from those adopted by NSF. OSIR has indicated taking a variety of actions in response to findings of misconduct in science in addition to the actions implemented by the research institutions (see Table 5.1). The OSIR's actions have included referral to the DHHS's OIG (when there have been findings of possibly criminal offenses), use of PHS sanctions (such as repayment of funds or debarment), and other institutional penalties (such as "letters of admonishment to subjects or institutions, a requirement that the employing institution send letters of reprimand to the subjects, and a requirement that the subjects of an investigation send letters of apology to the informant" (DHHS, 1991b, p. 6).

Remedial Actions

Some misconduct investigations have revealed problems that fall short of the regulatory definitions of misconduct in science but are judged to warrant remedial actions. These problems include "scientific sloppiness, incompetence, poor laboratory management, and poor authorship practices" (DHHS, 1991b, p. 4). Failure to implement the remedial action can result in a loss of future funding or other institutional penalties. Local institutions may also take remedial actions (such as withdrawing a research proposal), even if an inquiry results in a finding of no misconduct and no further investigation is conducted.

Faculty Participation in Misconduct Investigations

A particular problem arises when a government agency undertakes a review of an investigation that has been completed by a university. The university investigation is often undertaken by members of the research community who are requested by university officials to examine a matter that is usually complex and contentious. They are expected to do so to the best of their abilities.

When a government agency decides to review the university investigation, the agency makes clear that a potential for a conflict of interest exists between the university that commissioned the investigation and the individuals who undertook the investigation. Faculty members who participate in misconduct investigations can find, unexpectedly, that they themselves are subjects of the government agency review. If the risk and consequences to the university investigators from the subsequent agency review are significant, universities face the possibility that qualified university members will refuse to serve on committees that are formed to carry out an inquiry or investigation of alleged misconduct in science. This would be very unfortunate because it would serve to exclude those who may have the best understanding of the context in which the alleged misconduct took place.

The Role of Whistle-blowers

Individuals who bring soundly based allegations of misconduct in science to the attention of research institutions or government agencies perform an important function. The act of charging a colleague with inappropriate behavior requires both courage and the strong conviction that the observed behavior is wrong. Many research institutions are able to respond immediately to reports of suspected misconduct, and in these cases, the individual who originated the complaint is not required to take further action. On some occasions, however, individuals who initially disclosed misconduct have become the targets of investigation or retaliation, especially if the accused person holds a position of power or authority in the research institution. Many whistle-blowers have reported having experienced professional discrimination and economic loss as a result of their actions.[37] These experiences can discourage others from reporting misconduct in science.

Providing protections for whistle-blowers is difficult because the reprisals that may be taken against them can be subtle and indirect. A researcher's reputation, especially in the early stages of career development, depends greatly not only on scientific and technical achievement, but also on positive recommendations from collaborators and senior figures who can provide access to research resources. It is also difficult to assure job protections in a research enterprise that is often characterized by temporary and collaborative research assignments.

Once the whistle-blower has made an allegation of misconduct to an appropriate official, he or she is usually not a direct party to the misconduct inquiry or investigation. The whistle-blower may or may not serve as a witness or provide documentation for the charges. A whistle-blower who is dissatisfied with the adequacy of an institutional investigation may risk his or her scientific career by presenting an accusation to governmental authorities or the press. Some whistle-blowers have suffered serious harm even when their claims were correct.[38]

Furthermore, the initiation of a formal complaint of misconduct in science may result from a lengthy sequence of informal discussions and disputes between the complainant, the subject of the allegation, and other colleagues. Thus, in the human dimension, issues associated with the resolution of allegations of misconduct can be quite problematic. The stress and personal animosities of these cases can have a serious impact on the morale of an entire research group.

The Problem of False Accusations

When allegations of misconduct in science are misguided or malicious, the target of the claim also can suffer serious harm. The need to support those who report misconduct in science therefore must be balanced against the damage that can be done to honest scientists by false or malicious allegations. It cannot be assumed that all who bring allegations of misconduct in science are acting in good conscience. Professional rivalries, personal conflicts, or other complicating factors may stimulate false or malicious allegations, although the panel has not seen evidence of such allegations in the public record. Individuals need to be accountable for a complaint of misconduct in science, and appropriate documentation should be provided at the time of an initial allegation to justify the investment of institutional time and resources necessary to review the complaint.

ADDITIONAL FINDINGS AND CONCLUSIONS

Current Situation

The panel recognizes that the complexities of misconduct-in-science cases and their disposition are only beginning to emerge. Several factors can inhibit vigorous pursuit of misconduct-in-science allegations: concerns about individual reputations and the potential loss of institutional prestige, the lack of explicit channels for raising concerns about misconduct in science, confusion about the distinction between inquiries and investigations in misconduct-in-science procedures, the legal liability of institutions or of participants in the investigatory process, uncertainty about the legal standing of records of institutional investigations, ambiguity about the level of documentation that is sufficient to initiate or terminate inquiries or investigations, confusion about the level of evidence that is necessary to sustain findings of misconduct in science, and uncertainty about appropriate forms of disclosure of findings and sanctions. Consideration of such matters by each research institution *prior to* the treatment of an allegation or incident of misconduct in science would improve the process.

The panel is aware of the inherent difficulty posed by asking research institutions to investigate allegations of misconduct in science that involve their own members, especially when those members hold prominent positions of prestige and respect. Internal investigations must demonstrate a fundamental commitment to independence and objectivity to ensure their credibility and success, and may be enhanced by the participation of members from outside the affected organization. The objectivity of misconduct-in-science investigations relies heavily on the credibility of the process used to arrive at findings and recommendations. To maintain the privilege of self-regulation, research institutions must exercise vigilance and diligence in examining the conduct of their own members.

Current ad hoc efforts to foster dialogue about misconduct in science, other misconduct, and questionable research practices between research institutions and government agencies have raised many questions about appropriate roles, procedural flaws, and adequate resources in addressing these factors. There is a growing expectation that confirmed findings of misconduct in science should be reported to all individuals and institutions who might be affected.

The panel believes that it is important at this time to preserve institutional flexibility and discretion in developing and applying policies and procedures to address misconduct in science, but it is also important to clarify the basic criteria that will be used by faculty, administrative and governmental officials, and society as a whole in evaluating institutional methods for handling allegations of misconduct in science. It is necessary to include essential elements of fairness, objectivity, openness, and confidentiality in the investigations of alleged misconduct, and to reconcile competing interests, not only in principle but also in practice.

In considering protections for whistle-blowers in misconduct proceedings, the panel formulated three fundamental principles:

1. Whistle-blowers should be assured that their claims will be taken seriously and will receive full and fair consideration by responsible officials.

2. In some cases, whistle-blowers may seek anonymity as a protection against reprisals and discrimination, especially while an inquiry or investigation is in the early stages of development. While such anonymity may be desirable, there may be practical constraints in assuring confidentiality in a highly specialized research area or in a small research team.

3. When reprisals against whistle-blowers are discovered, the responsible individuals should be punished in accord with the severity of the reprisals. The standards for examining complaints about possible reprisals in the academic environment should be consistent with those developed for federal employees under the Whistleblower Protection Act of 1989.[39]

Balancing Accountability and the Need for Intellectual Freedom

In the wake of procedural and policy reforms in response to incidents of misconduct in science, representatives from the academic and scientific community have raised concerns about the long-term or unintended effects that might result from institutional or governmental intrusions into the research environment.[40] Aggressive efforts to control research practices, if carried to an extreme, can damage the research enterprise. Balance is required. Inflexible rules or requirements can increase the time and effort necessary to conduct research, can discourage creative individuals from pursuing research careers, can decrease innovation, and can in some instances make the research process impossible. Governmental or regulatory efforts to define "correct" research conduct or analytical practices can do fundamental harm to research activities if such efforts encourage orthodoxy and rigidity and inhibit novel or creative research practices.

However, the panel concludes that allegations and incidents of misconduct in science require a vigorous institutional response and that the methods used by research institutions and government to address allegations of misconduct in science need improvement. Research institutions sometimes require advice or assistance in addressing allegations of misconduct in science because of the complexities of these cases or because their faculty or administrators are reluctant to address in a systematic manner complaints or suspicions about possible misconduct in science. Research institutions have not developed mechanisms for broad exchange of information and experience in resolving difficult cases and consequently lack opportunities for learning from each other. On several occasions, institutional officials have waited for direction from government agencies before clarifying their own procedures for handling allegations of misconduct in science.

Need for an Independent Body
as an Additional Resource

In considering responses by research universities, government, and the courts to incidents of misconduct in science, the panel concludes that additional resources are required to strengthen the processes and procedures used for handling and resolving allegations of misconduct in science. This conclusion is derived from the following findings:

• *First*, the panel believes that some research institutions have been slow to respond to and to pursue alleged cases of misconduct in science within their own communities. The panel concludes that an independent organization could be an effective resource to assist individual research institutions by sharing knowledge of "best practice" among the community.

• *Second*, the research community has not been effective in responding to criticism about its record in handling allegations of misconduct in science. As a result, firsthand experiences in resolving problems of fairness, responsiveness, and accuracy in misconduct proceedings are often not systematically analyzed or disseminated to improve the resources and methods used by research institutions in handling allegations of misconduct in science.

The panel believes that a knowledgeable and credible voice is needed in the debate about the effectiveness of the scientific community in meeting the public interest. This perspective should not be tainted by the accusation that a voice is protecting the interest of a particular research institution or individual under scrutiny.

• *Third*, the panel notes that several government agencies, notably the NSF and the PHS, have established offices for dealing with allegations of misconduct in science by their grantees. The panel is concerned about the scope of current government definitions of misconduct in science, the ability of government offices to handle allegations of misconduct in science effectively, and the possibility that the system established to handle misconduct in science could stray into matters that lie more appropriately in the domain of the scientific community (such as the detection of scientific error, the development of scientific methodologies, or the rejection or confirmation of new theories of scientific phenomena).

The panel concludes that the scientific community, Congress, federal authorities, and the public should have a single, independent body available to comment with knowledge and credibility on how working scientists, research institutions, and government agencies are progressing at meeting the common objective of ensuring integrity in the research process and responding vigorously and fairly to alleged misconduct. An independent organization could perform this important function if constituted and operated in an appropriate way. Further discussion and pertinent details are given in Chapter 7.

NOTES

1. See, for example, the discussion regarding the allegations against Franz Moewus as presented in Sapp (1990).

2. The survey consisted of a stratified random sample of 30 institutions with 100 or more grants, 31 with 10 to 99 grants, inclusive, and 28 institutions with fewer than 10 grants.

3. See Department of Health and Human Services (1989a). See also National Science Foundation (1987, 1991b).

4. See, for example, the report of a conference on misconduct in science sponsored by the American Association for the Advancement of Science, the American Bar Association, the National Conference of Lawyers and Scientists and the DHHS's Office of Scientific Integrity Review (AAAS, 1991b).

5. The NSF has taken the position that although an inquiry can produce a finding of no misconduct, an investigation is necessary to establish misconduct. Personal communication, OIG, NSF, February 1, 1991.

6. See Department of Health and Human Services (1991a).

7. See, for example, the "Dear Colleague" letter issued by the NSF's OIG on August 16, 1991.

8. See Greene et al. (1985). The survey was sent to 747 institutions, and 493 (66 percent) responded.

9. See Department of Health and Human Services (1991b). Also, National Science Foundation (1990b, 1991a).

10. Examples of misconduct policies and procedures from the institutions discussed in this section are included in Volume II of this report.

11. See, for example, the discussion in the DHHS's OIG report (DHHS, 1989d), which notes that although all "large grantee institutions considered [misconduct] investigations their responsibility, . . . only 54 percent of the small institutions shared this view, and most of these institutions would support a more active NIH role in investigating allegations" (p. 11).

12. For a full discussion of some procedural complexities involved in academic investigations of misconduct-in-science allegations, see the proceedings of a series of workshops sponsored by the National Conference of Lawyers and Scientists (AAAS-ABA, 1989).

13. See the statement by Rep. John Dingell in U.S. Congress (1989b): "The apparent unwillingness on the part of the scientific community to deal promptly and effectively with allegations of misconduct is unfair to both the accuser and to the accused" (p. 1). See also Weiss (1991b) and the commentary in Dong (1991).

14. For an informative discussion, see Andersen (1988).

15. See, for example, Department of Health and Human Services (1991a).

16. Public Law 100-504 (102 Stat. 2515 [1988]) established Offices of Inspector General in a number of departments and smaller agencies, including NSF. In compliance

with the legislation, the National Science Board established the NSF's Office of Inspector General on February 10, 1989.

17. See, for example, the minutes of the meetings of the OSIR Advisory Committee held July 15, 1991, and November 17, 1991 (DHHS, 1991d).

18. See, for example, Hallum and Hadley (1990).

19. See Hamilton (1991b) and Weiser, B. 1991. "NIH alleges misconduct by Georgetown scientist." *Washington Post* (March 22):A1.

20. Personal communication, OSIR, June 19, 1991.

21. One particularly troubling issue in the investigation of allegations of misconduct in science concerns the nature of the review or hearing that should be provided. Several principles of fairness, confidentiality, and completeness may come into conflict during this stage. For example, should the accused be allowed to cross-examine witnesses, including the complainant who filed the initial allegation? Although a formal hearing may be appropriate when specific and serious penalties have been proposed, there is no consensus that a subject is entitled to review testimony or to cross-examine witnesses during the fact-gathering process designed to provide evidence to substantiate or dismiss charges of misconduct in science. In contrast, there is general agreement that the subject should be given access to the draft investigative report for rebuttals, modifications, or other amendments prior to the formulation of specific charges or a dismissal of the complaint. Some institutions have also provided access to the draft report to significant witnesses, including the initial complainant, although this is not customary.

22. See 18 U.S.C. Sections 287, 1001 (1988); *United States* v. *Breuning*, No. K88-0135 (D. Md., Nov. 10, 1988).

23. For example, after a guilty plea on two counts of making false statements to the government on grant applications (issued in exchange for dropping a charge of obstruction of the government's investigation of his conduct), Stephen E. Breuning was sentenced in 1988 to 5 years probation, 2 months in a half-way house, and 250 hours of community service. He was ordered to repay $11,352 of salary for the time covered by his fraudulent research and to conduct no psychological research during the period of his probation. See Frankel (1988).

24. *Abbs* v. *Sullivan*, 756 F. Supp. 1172 (W.D. Wis. 1990).

25. "Nor shall any person . . . be deprived of life, liberty, or property, without due process of law." U.S. Constitution, Fifth Amendment. State governmental action is similarly limited by the due process clause of the Fourteenth Amendment.

26. *Fuentes* v. *Shevin*, 407 U.S. 67, 97 (1972).

27. See pp. 666-67 in Tribe (1988). See also *Joint Anti-Fascist Refugee Committee* v. *McGrath*, 341 U.S. 123, 168 (1951) (Frankfurter, J., concurring; procedural safeguards give the accused "the right to be heard before being condemned to suffer grievous loss").

28. See *Perry* v. *Sinderman*, 408 U.S. 593, 601-03 (1972) (plaintiff, an untenured instructor, could have liberty or property interest in continued employment); see also *Ferguson* v. *Thomas*, 430 F.2d 852, 856 (5th Cir. 1970) ("a college can create an obligation as between itself and an instructor where none might otherwise exist. . . if it adopts regulations or standards of practice governing non-tenured employees which create an expectation of reemployment") and *Board of Regents* v. *Roth*, 408 U.S. 564, 578 (1972) (non-tenured instructor had no right to a hearing before the university decided not to renew his contract, absent custom or mutual agreement that his employment would be renewed).

29. *Paul* v. *Davis*, 424 U.S. 693, 701 (1976) ("reputation alone, apart from some more tangible interests such as employment, is [n]either 'liberty' [n]or 'property' by itself

sufficient to invoke the procedural protection of the Due Process Clause"). But see *Wisconsin* v. *Constantineau*, 400 U.S. 433, 437-39 (1971), and *Jenkins* v. *McKeithen*, 395 U.S. 411, 426-31 (1969).

30. See, for example, *Matthews* v. *Eldridge*, 424 U.S. 319, 335 (1976).

31. The elements of due process, any or all of which may be required before an individual can be deprived of a particular liberty or property interest, include (1) adequate notice of expected conduct; (2) adequate notice of the charges; (3) a timely hearing; (4) a neutral decision maker; (5) an opportunity to make an oral presentation to the decision maker; (6) an opportunity to present evidence or witnesses to the decision maker; (7) a chance to confront and cross-examine witnesses or evidence to be used against the accused; (8) the right to have an adviser involved to assist in the presentation of the individual's case to the decision maker; and (9) a decision based on the record with a statement of reasons for the decision.

Not all of these elements must be present in every hearing. To the contrary, only the more serious deprivations of liberty or property interests by the state require extensive procedural safeguards.

See generally, pp. 706-18 in Tribe (1988) and pp. 555-56 in Nowak et al. (1983). See also Mishkin (1988).

32. In this discussion the panel uses the term "hearing" to refer to the mechanism used to investigate complaints of misconduct in science. The hearing may or may not involve sessions in which the subject of a misconduct investigation may hear testimony by or cross-examine witnesses.

33. The issue of the involvement of attorneys in the investigation of misconduct in science is a vexing problem. Some universities believe that attorneys should not participate in the university investigatory process because their involvement may lead to an adversarial spirit that is not consistent with the academic environment. Other universities allow those accused of misconduct to be represented by attorneys in the misconduct investigation. In such cases, the investigative panel may also request the university to supply its legal counsel for the panel's assistance.

34. 45 C.F.R. Section 689.2(b) (1991).

35. 45 C.F.R. Sections 620.314(c), 689.2(d) (1991).

36. 45 C.F.R. Section 689.2(a) (1991).

37. See, for example, the accounts published in Westin (1981) and in Glazer and Glazer (1990).

38. Swazey and Scher (1981) and Glazer and Glazer (1990). See also Hollis (1987), Jacobstein (1987), and Sprague (1987).

39. Public Law 101-12 (103 Stat. 16 [1989]).

40. See, for example, testimony by academic officials and scientists in hearings on maintaining the integrity of scientific research convened by the House Committee on Science, Space, and Technology (U.S. Congress, 1990b).

6

Steps to Encourage Responsible Research Practices

ACKNOWLEDGING RESPONSIBILITY AND TAKING ACTION

The size, complexity, and diversity of research efforts, among other factors, contribute to excellence in a changing and competitive scientific research environment. However, these same features can provide opportunities for misconduct in science, questionable research practices, and other misconduct. Individual scientists bear the primary responsibility for the conduct of their research, but local research institutions and sponsoring organizations also have responsibilities, in addition to implementing fair, sound, and well-defined mechanisms to investigate allegations of misconduct in science. Research institutions strive to provide a climate that encourages responsible practices and discourages questionable research practices. The challenge to research institutions is to aid faculty in establishing effective systems of values and social controls, to provide individuals with opportunities and incentives to develop and implement these systems, and to safeguard the traditions that foster scientific creativity.

Institutional efforts to encourage responsible research practices have been stimulated by the following factors:

• Growth and diversification of research, creating situations likely to be sources of increasing disputes about appropriate forms of research behavior. In addition to relying on traditional methods of individual instruction and professional example, research institutions are seeking more explicit ways to aid their members' efforts to discriminate between acceptable and unacceptable research practices.[1]

• Recognition that many types of research practices that do not constitute misconduct in science are nevertheless questionable and fall well short of responsible research behavior. Scientists and the public in general are likely to grow dissatisfied with self-serving research practices that erode communal values and standards.

• Regulations requiring institutions that receive research funds from the Public Health Service (PHS) to establish an environment that discourages misconduct in science.[2] In addition, applicants for biomedical training grants funded by the National Institutes of Health (NIH) and by the Alcohol, Drug Abuse, and Mental Health Administration (ADAMHA) must now demonstrate that they provide instruction in the "responsible conduct of research" in their training programs.[3]

• Belief that sustained efforts by the research community to strengthen the accountability and integrity of the research environment may obviate the need for additional federal intervention.

Some research institutions have sought to develop educational programs or guidelines intended to foster responsible research practices. The effectiveness, desirability, and need for such programs and guidelines have been debated and discussed within the research community. Although many advocate expansion of the research institution's role in fostering responsible research practices, others— often individual faculty members—have expressed caution based on the following assumptions:

• Institutional efforts designed to foster integrity in the research environment may be misinterpreted as an admission that the system is not working well or that faculty are not exercising their responsibilities.

• Institution-wide programs designed to encourage responsible research practices may weaken individual and departmental efforts to achieve the same goals. Institutional programs may all too easily intrude on and replace the more personal—and possibly more effective—efforts of individual scientists who regard the fostering of scientific responsibility as a professional obligation.

• Self-imposed institutional guidelines or educational programs may encourage government to utilize this mechanism for inappropriate oversight.

INTEGRATING ETHICS INTO THE
EDUCATION OF SCIENTISTS

The entire scientific community bears a responsibility for ensuring that the customs, traditions, and ethical standards that guide responsible research practices are systematically communicated to research scientists and trainees. As mentors, practicing scientists often impart these values to their students and associates, who thus can learn through direct guidance and also by example the customs of responsible research practice. But formal or explicit definition of standards governing the responsible conduct of research is infrequent.

Benefits of Education in Ethics

Although data reviewed in Chapter 4 indicate that young investigators or students are perhaps less likely than older researchers to engage in misconduct in science—in fact, many cases of misconduct have involved senior researchers—early education can be a primary means of instilling responsible practices. Studies in the literature on ethics education suggest that ethical development is not complete or fixed by the time students go to graduate school (Rest, 1988). Thus, although ethics education alone is unlikely to change individual moral character, teaching ethics in a professional setting can foster awareness and can reinforce the importance of actions that constitute appropriate behavior in the conduct of research.

For example, informal and formal discussions of genuine ethical problems that arise in the research environment—such as the allocation of credit for a collaborative effort that involves specialized contributions—can teach both students and faculty about the significance and consequences of alternate responses to difficult situations. Moreover, the public nature of educational discussions can create a climate that may discourage individuals from engaging in questionable practices, as students and colleagues examine the potential harm that such practices can cause. Regularly held graduate seminars, faculty colloquia, and informal discussions in the laboratory and the classroom can also provide opportunities to test perceptions of observed practices against the expected norms of science, can help all members of the research community to define and clarify the fundamental norms that guide research practice, can ameliorate misunderstandings that could escalate into unfounded accusations, and can stimulate open and frank consideration of conflicting values. Exploring a case of poor authorship practices in the context of a classroom discussion of questionable research practices, for example, might be less threatening to a concerned student than approaching an institutional officer or departmental superior.

Finally, education in research ethics can help all involved in the research process to become informed participants in the self-governance of the scientific community. The ideal of informed participation is as important for members of the scientific community as it is for citizens of the larger political community.

Approaches to Teaching Ethics

Various approaches can be adopted in teaching research ethics. One involves examining the special obligations scientists have by virtue of their expert knowledge and profession and clarifying how practices and standards may differ among disciplines or among institutions. Instruction based on this approach could include discussions of standards of good practice, misconduct in science, questionable research practices, and other misconduct. Specific topics that should be addressed include the following:

• The necessity of honesty, skepticism, error correction, and verification in science;

• Principles of data selection, management, and storage, including rights and responsibilities with respect to sharing and granting access to research data, and the special status of data that support published findings;

• Publication practices, including the importance of timely and appropriate release of significant research findings and the harm that can result from premature or fragmentary publication of results or from publication in multiple forms;

• Authorship practices, particularly criteria for and obligations of authorship and the proper allocation of credit for specialized contributions; and

• Training and mentorship practices, including the responsibilities of supervision and the principles that guide collaboration between senior and junior personnel.

Some honorary and professional societies have prepared educational materials to encourage discussions of such topics. The National Academy of Sciences, for example, has published *On Being a Scientist* (NAS, 1989), an essay written to instruct graduate students in the values and practices of scientists, and Sigma Xi has made its educational essay on ethics and science, *Honor in Science* (Sigma Xi, 1986), widely available. *Scientific Freedom and Responsibility*, prepared by Harvard biologist John T. Edsall (1975) for the American Association for the Advancement of Science, was an early report that still would enhance the quality of current discussions about appropriate behavior in science.

A second approach to teaching ethics focuses on examining laws, institutional policies, and professional standards that guide certain fields of activity (such as the use of human subjects in biomedical, social, or behavioral research or the use of hazardous materials in the natural sciences). Such an approach can clarify the justification for adopting particular rules and also can explain the context and some of the abuses and value conflicts that spurred the development of specific rules and standards. Discussions of institutional policies should be explicit about appropriate channels for raising concerns when one witnesses misconduct in science, questionable research practices, or other misconduct. Such discussions may help prevent conflicts that can result from poor communication or poorly understood expectations about what behaviors constitute misconduct in science or questionable research practices.

A third approach involves going beyond laboratory and classroom discussions of responsibility in research to consider specific ethical questions in the broader context of competing rights and obligations in the research community. University-wide forums can provide opportunities to discuss authorship, communication, and data-handling practices that may both educate faculty and students and allow comparison of different disciplinary practices. Research institutions could also provide funds to graduate students, interns, and other junior scientists to organize discussion sessions and to prepare case studies to highlight current ethical dilemmas. Such forums and sessions could also facilitate interdisciplinary discussions of the philosophy, history, and social studies of science that bear on scientific conduct.

Experience gained in teaching engineering ethics and biomedical ethics suggests that the following principles can contribute to the success of ethical discussions as they are integrated into scientific or engineering programs:

• More than generalities should be taught. Specific examples, preferably local case histories, are the preferred way to provide guidance on matters important in the profession.

• Education must aim at influencing behavior. Professional training cannot assure that people will make correct moral judgments, but it can provide the opportunity to learn from experts who can explain the reasoning behind certain moral judgments or professional practices.

- Perspectives gained from looking beyond science itself are valuable in examining ethical issues in the research environment. Studies in ethics, moral philosophy, history, and the social studies of science can contribute to a broader outlook that can aid in rethinking controversial issues and establishing values in research.

- If properly structured, topics and teaching materials related to ethics in science and research can be intellectually stimulating for students and faculty. Such topics can be taught in dedicated courses or included in courses within the broad curriculum. The panel's discussions with students and faculty indicate that both approaches are desired by the larger community.

As noted above, universities that have applied for NIH or ADAMHA training grants must develop educational programs to foster broad discussions of responsible research practices. The NIH has convened several workshops to examine the strengths and limitations of various approaches to fulfilling the training grant requirement.[4]

Some departments and universities have sponsored forums and seminars that offer students the benefit of learning from watching faculty grapple seriously with issues involving responsible practice. Real or hypothetical case studies are also useful devices for examining selected research practices. Relevant instruction and the message that responsibility in research is to be taken seriously can also be given in orientation programs for new graduate students, postdoctoral fellows, and faculty.

Interdisciplinary training workshops may improve the quality of instruction and curriculum materials for teaching ethics in scientific research. After a period of years, and when a significant number of schools have developed curricula on research ethics, it could be useful to review and to improve as necessary the quality of teaching and of the curriculum materials used for instruction in research ethics.[5] Such a review could draw on the expertise and judgments of a consensus panel representing those engaged in ethics instruction as well as those who are respected scientists in the fields under study.

CONSIDERING GUIDELINES FOR RESPONSIBLE RESEARCH PRACTICES

Current Means for Providing Guidance

Even though most research institutions do not have written guidelines for the conduct of research, their faculty usually act individually and informally to encourage responsible research practices.[6] In addition, most universities have (1) general codes of academic conduct or honor codes that apply broadly to faculty, administrators, staff, and students[7] and that provide for disciplinary action by the institution in the event of serious violations and (2) written policies dealing with specific issues in the research environment, such as conflict of interest, intellectual property rights, use of humans and animals in experimentation, and computer use.[8] Most academic institutions that conduct significant amounts of research have also adopted policies and procedures for handling allegations of misconduct in science.[9]

The normative rules and monitoring requirements scattered throughout university policies and documents relating to science and engineering research are an important first step for promoting the responsible conduct of research. In defining what is illegal, unethical, and irresponsible, these rules suggest what is legal, ethical, and responsible. For example, the University of Maryland policy on misconduct defines "improprieties of authorship" as "improper assignment of credit, such as excluding others; misrepresentation of the same material as original in more than one publication; inclusion of individuals as authors who have not made a definite contribution to the work published; or submission of multi-authored publications without the concurrence of all authors" (University of Maryland at Baltimore, 1989, p. 2). This statement could be interpreted as a guideline for responsible behavior in research, since it encourages the proper assignment of credit for research performance and urges authors to include the names of co-authors only with their permission.

Therefore, although most research institutions do not have comprehensive codes of conduct for science and engineering research, they do provide ethical and policy guidance to researchers. If these policies are considered along with the various federal regulations, statements of professional societies about professional conduct in research, and other literature prepared by professional and scientific societies (such as the National Academy of Sciences' essay *On Being a Scientist* and Sigma Xi's essay *Honor in Science*), the total package provides a strong foundation for describing what is responsible and irresponsible in the conduct of research.

However, the currently existing set of normative rules designed to foster responsibility in science has limitations. Research policies are often disjointed and piecemeal, they may be administered by different academic units, and they may vary substantially among institutions. It is difficult for researchers to comprehend and consider all the legal and professional responsibilities raised by modern science and engineering. Yet integrating rules and resolving contradictions are often left to the individual, who is expected to read through three, four, or more separate policies to determine what is proper.

The use of examples or case studies that deal with difficult rather than obvious issues is a valuable method of interpreting and explaining policy statements about normative or ideal conduct. Few doubt that manufacturing data or forging experimental results is wrong. It may be less clear, however, how preliminary results should be presented in grant applications, when "enough data" are needed to give confidence that a project will succeed but "enough work" remains to be done to justify the grant award.

Most normative rules provide important general principles but leave significant questions unanswered. This void has prompted some universities to take additional steps to foster responsible conduct in research, such as developing guidelines for the conduct of research.

Scope and Purpose of Institutional Guidelines for the Conduct of Research

By "guidelines for the conduct of research," the panel means institutional policies that address practices such as those related to data management (including data collection, storage, retention, and accessibility), publication (including authorship policies), peer review and refereeing, and training and mentorship. Some institutions have guidelines that focus on a single topic, such as authorship, whereas others adopt a more comprehensive approach. The guidelines may be voluntary or compulsory, and they are administered through a variety of organizational units.

Several major research institutions, such as the National Institutes of Health (for its intramural research program), Harvard Medical School, Johns Hopkins University Medical School, and the University of Michigan Medical School, have formulated comprehensive guidelines for the conduct of research.[10] Nevertheless, comprehensive guidelines for research conduct are not common. One study of 133 medical institutions indicated that 17 (13 percent) had such guidelines and that 25 (19 percent) were considering developing guidelines, while 91 (68 percent) were not (Nobel, 1990).

Guidelines for the conduct of research differ from institutional policies that are designed to address misconduct in science or conflict of interest or that, in response to regulatory requirements, govern research involving human subjects, hazardous materials, or recombinant DNA.[11] Research conduct guidelines are intended to promote responsible conduct of research and, to the extent that questionable practices and misconduct in science are linked, to reduce the amount of misconduct in science. However, although there are positive advantages associated with the adoption of such guidelines, this approach, by itself, may not be effective in fostering responsible research conduct. The imposition of guidelines also carries certain risks and limitations in the collegial academic environment.

Benefits of Institutional Guidelines for the Conduct of Research

Research conduct guidelines represent an important, but not necessarily the best, means by which research institutions can demonstrate awareness of and support for principles of good research practice. Properly constructed and used, research guidelines can help articulate and strengthen the fundamental values of scientists, especially in an increasingly diverse and changing research environment.[12]

In principle, research conduct guidelines can help scientists to understand the criteria that should be considered in, for example, making decisions about how long research data should be retained. They can also help clarify for scientists and trainees what constitutes "good practice," although many research practices are, by virtue of their complexity, subject to varying interpretations.

Government agencies already require research institutions to play a stronger role in fostering responsible research practices, and it is possible that such guidelines may one day be required as a condition of governmental funding. Thus it may be wiser to have research conduct guidelines developed internally by faculty and research scientists who are most familiar with their own institutional research environment than to have them imposed by higher authorities to fulfill funding or regulatory requirements.

Guidelines may help inform members of a research institution about what constitutes questionable practices or misconduct in science in an academic research environment. For example, by issuing guidelines that state the criteria for authorship, universities can fulfill a due process obligation to provide notice to their members of the unacceptable authorship practices, such as plagiarism, that may constitute grounds for disciplinary actions.

Disadvantages of Institutional Guidelines for the Conduct of Research

Many scientists believe that research conduct guidelines are unnecessary and ineffective, and they point out that research practices are often too complex and too varied to be governed by a few general principles. Moreover, adopting such principles is a time-consuming process that requires the efforts of active researchers who are already burdened by other obligations. Some are concerned that focusing on guidelines diverts attention from the consideration of complex ethical issues and genuine dilemmas in the research environment.

The concept of research guidelines cuts against faculty autonomy and other values associated with academic freedom as ideals of the academic environment. In the past, steps that might restrict scholarly or scientific independence were taken only when there was clear evidence that inappropriate behaviors or hazardous situations might persist in the absence of institutional policies.

Although they themselves may not require new administrative procedures, guidelines may encourage implementation of rules and rigid regulations, as well as "cookbook" approaches to scientific endeavors. Some institutional officers are also concerned that guidelines that describe appropriate research conduct will expose them to additional litigation and administrative vexations (Nobel, 1990).

Many scientists believe that explicit guidelines can add little of substance to the material already included in publications such as *On Being A Scientist* and *Honor in Science*. Others think that guidelines may not be necessary because the norms of science are self-evident and because existing policies provide abundant advice to determine how to conduct research responsibly. Another drawback is the effort needed to bring to the attention of faculty and students any research conduct guidelines that have been adopted. If research conduct policies are not appropriately implemented, they can be viewed as empty gestures or "window dressing" that will serve little purpose.

Conclusions About Institutional Guidelines for the Conduct of Research

In considering the advantages and disadvantages of guidelines for research conduct, the panel concluded the following:

1. Guidelines that are relevant and appropriate to research may be widely disparate depending on the research field, the nature of the work, and other factors.

2. Written guidelines are unlikely to influence academic research behavior if they are imposed from above or from outside. The process of formulating guidelines may be extremely valuable for those who participate; however, efforts will need to be made to ensure that the final statements express the fundamental ideas and potential conflicts inherent in such guidelines.

3. To be effective, guidelines must be incorporated into the process of research and education and become an operational part of day-to-day activities. If faculty desire to develop guidelines for the conduct of research, such policies should be formulated by those who will be directly affected and should be adapted to specific research fields and protocols.

Institutional guidelines are likely to be less effective than ones formulated at the group or laboratory level. However, research institutions may wish to adopt an overarching set of general principles for their members to provide a common frame of reference. The panel recognizes that the formulation of written guidelines is an exacting task that requires substantial time and effort. Guidelines may help clarify the professional obligations of faculty and research staff, but the panel believes that the development of such guidelines should be left to the discretion and initiative of individual faculty and research institutions. In any case, care should be taken to avoid adopting constraints that could be damaging to the research process.

A FRAMEWORK OF SUBJECTS TO CONSIDER IN ENCOURAGING RESPONSIBLE RESEARCH PRACTICES

The panel has identified a set of subjects that should be considered in any efforts aimed at developing educational discussions or guidelines for the conduct of scientific research. This set of subjects is not meant to be comprehensive but rather to suggest particular topics and examples of "best scientific practice" that should be considered in formulating statements on research conduct. Examples of selected guidelines are paraphrased below to illustrate ways in which different institutions have addressed these topics.[13]

1. *Data management.* Acquisition and maintenance of research data should be addressed since they provide the foundation for scientific discovery and experimentation. Research data include detailed experimental protocols, primary data from laboratory instruments, and the procedures applied to reduce and analyze primary data.
   a. *Subjects to be addressed:*
   • Availability of data to scientific collaborators or supervisors
   • Retention of data for specified periods of time
   • Accessibility of data after publication
   b. *Examples of good practice:*
   (1) Research data, including the primary experimental results, should be retained for a sufficient period to allow analysis and repetition by others of published material from those data. In

some fields, 5 or 7 years are specified as the minimum period of retention, but this may vary under different circumstances. (NIH, 1990)

(2) Custody of all original primary laboratory data must be retained by the unit in which they are generated. An investigator may make copies of the primary data for his/her own use. (Harvard University Faculty of Medicine, 1988)

(3) All primary data are to be entered into a notebook provided by the institute for this purpose. The investigator is responsible for all data entries. The notebook will contain lined, numbered pages; no pages are to be removed or made illegible. Entries must be dated and signed. (Dana–Farber Cancer Institute, 1987)

(4) All data, even from observations and experiments not leading directly to publication, should be treated comparably. Research data should always be immediately available to scientific collaborators and supervisors for review. In collaborative projects involving different units, all investigators should know the status of all contributing data and have direct access to them. (NIH, 1990)

2. *Publication practices.* Science is a cumulative activity in which each scientist builds on the work of others. Publication of results is an integral and essential component of research because it enables others to gain access to each scientist's contribution.

a. *Subjects to be addressed:*
- Methods of publication and disclosure of new findings
- Correction of errors and retraction of published findings
- Treatment of fragmentary results of a scientific investigation
- Multiple publications of same or similar findings
- Completeness of publication so that repetition and evaluation are feasible

b. *Examples of points to be kept in mind:*
(1) Certain practices make it difficult for reviewer and reader to follow a complete experimental sequence. Among these are the premature publication of data without adequate tests of reproducibility or assessments of significance, the publication of fragments of a study, and the submission of multiple similar abstracts or manuscripts differing only slightly in content. In such circumstances, if any of the work is questioned, it is difficult to determine whether the research was done accurately, the methods were described properly, the statistical analyses were adequate, or appropriate conclusions were drawn. In-

vestigators should review each proposed manuscript with these principles in mind. (Johns Hopkins University School of Medicine, 1990)

(2) In a publication, all data pertinent to the project should be reported, whether supportive or unsupportive of the thesis or conclusions. Except for review articles, publishing the same material in more than one paper should be avoided. Unnecessary fragmentation of a complete body of work into separate publications should be avoided. Prior work in the field should be referenced appropriately. (University of Michigan Medical School, 1989)

3. *Authorship.* Authorship and allocation of credit are primary benchmarks of achievement and rewards for scientists.
   a. *Subjects to be addressed:*
   • Criteria for authorship and identification of contributors
   • Order of listing of authors
   • Responsibility for authorship: collective and individual
   b. *Examples of good practice:*
   (1) For each individual the privilege of authorship should be based on a significant contribution to the conceptualization, design, execution, and/or interpretation of the research study, as well as a willingness to take responsibility for the defense of the study should the need arise. In contrast, other individuals who participate in part of a study may more appropriately be acknowledged as having contributed certain advice, reagents, analyses, patient material, support, and so on, but not be listed as authors. It is expected that such distinctions will be increasingly important in the future and should be explicitly considered more frequently now. (NIH, 1990)
   (2) Criteria for authorship of a manuscript should be determined and announced by each department or research unit. The [Harvard University Faculty] committee considers the only reasonable criterion to be that the co-author has made a significant intellectual or practical contribution. The concept of "honorary authorship" is deplorable. The first author should assure the head of each research unit or department chairperson that s/he has reviewed all the primary data on which the report is based and provide a brief description of the role of each co-author. (Harvard University Faculty of Medicine, 1988)

4. *Peer review.* Peer review is used to guide decisions on the funding of research and on the publication of research results. It is an essential component of the scientific research process.

a. *Subjects to be addressed:*
   - Responsibility to participate in the peer review process
   - Considerations of confidentiality and proprietary interests in peer review
   - Conflicts of interest and need for disclosure in peer review of competitive proposals
   - Objectivity of peer reviews; inclusion of nonpublic information

b. *Examples of good practice:*
   (1) It is important that reviewers and readers be informed of the sponsorship of research projects in order that they may be alert to possible bias in the research arising from a sponsor's financial interest in the results. (Johns Hopkins University School of Medicine, 1990)

   (2) The reviewer has the responsibility for preserving the integrity of the review process. In receiving a manuscript or a grant proposal, he is entrusted with privileged information that is unavailable to anyone outside of the laboratory of the submitting scientist(s). It is of obvious importance for the reviewer not to make use of information gained in the review for his own purposes until it is published or, prior to that, only by consent of the author. . . . The contents of a work under review should not be distributed to other colleagues. There are certain exceptions to this general rule, however. For example, it should be permissible to discuss parts or even all of a submitted work with trusted colleagues to obtain a second opinion in instances when the reviewer is unfamiliar with the methodology or considers the author to be mistaken. (University of Michigan Medical School, 1989)

5. *Training and supervision.* Scientists in universities accept the obligation to pass along knowledge and skills to the next generation of scientists.

a. *Subjects to be addressed:*
   - Assignment of mentors to students
   - Availability of mentors and appropriate forms of supervision
   - Degree of independence and responsibility for students and postdoctoral trainees
   - Types of duties assignable to students by mentors and supervisors
   - Appraisals and communication of student and trainee performance

b. *Examples of good practice:*
   (1) Each trainee should have a designated primary scientific

mentor. . . . The mentor has the responsibility to supervise the trainee's progress closely and to interact personally with the trainee on a regular basis in such a way as to make the training experience a meaningful one. . . . Mentors should limit the number of trainees in their laboratory to the number for whom they can provide an appropriate research experience. (NIH, 1990)

(2) The preceptor should provide each new investigator (whether student, postdoctoral fellow, or junior faculty) with applicable government and institutional requirements for conduct of studies involving healthy volunteers or patients, animals, radioactive or other hazardous substances, and recombinant DNA. (Harvard University Medical School Faculty, 1988)

(3) The preceptor should supervise the design of experiments and the processes of acquiring, recording, examining, interpreting, and storing data. A preceptor who limits his/her role to the editing of manuscripts does not provide adequate supervision. (Harvard University Medical School Faculty, 1988)

DISCOURAGING QUESTIONABLE RESEARCH PRACTICES

Many scientists and students do not believe that they will experience situations involving fabrication, falsification, or plagiarism. Yet, sometimes on a daily basis, they face situations that require ethical judgment and professional guidance. Students and young scientific investigators, in particular, may experience questionable practices, sometimes encouraged by their mentors, that cause them to question the fundamental values that should guide the responsible conduct of research. Requests for co-authorship in exchange for the preparation of unique samples or reagents for complicated experiments, for example, can be problematic for inexperienced as well as senior investigators. Rules and regulations often do not provide appropriate guidance for resolving such problems, which nevertheless cannot simply be tolerated or ignored. It is important to recognize that junior investigators may be particularly at risk in failing to distinguish, or prevent, unacceptable research practices.

Although questionable research practices are not appropriate for treatment as incidents of misconduct in science, they require the sustained attention of scientists and responses by institutional officers when there is general agreement that specific practices are not to be tolerated. The panel points out that the methods for addressing questionable research practices should be different from those for handling misconduct in science or other misconduct. Attention to questionable research practices should be rationalized

and regularized to encourage responsible research behaviors and to discourage questionable practices.

These questionable practices might include, for example, not giving colleagues access to data or research materials; failing to retain, for a reasonable period, data or research materials that support reported findings; designating as an author one who has made no significant contribution to a paper, as well as failing to acknowledge as an author an individual who has made a significant contribution to the work reported in a paper; or exploiting graduate students.

Recognizing that specific approaches may have important limitations, the panel nevertheless concludes that it is essential for scientists and research institutions to exercise a stronger role in providing an environment that encourages responsible research practices and also discourages misconduct in science.

In considering different approaches to dealing with questionable research practices, the panel concluded that questionable practices are best discouraged through (1) the effective use of peer review and the system of appointments, evaluations, and other rewards in the research environment and (2) educational programs that emphasize responsible behavior in the research environment. Such approaches build on the strengths of self-regulation, rely on those who are most knowledgeable about the intricacies of the scientific research process to maintain the quality of the research environment, and preserve the diverse disciplinary traditions that foster integrity in the research process. By encouraging the development of educational programs that emphasize responsible research behavior, the panel seeks to foster more deliberate and informed communication, discussion, criticism, and reflection of the basic values that guide scientific practices and judgments.

The role of government should be confined to one of providing oversight of institutional efforts to handle and prevent episodes of misconduct in science. Government should not seek to regulate questionable research practices.

NOTES

1. See, for example, the report of the Massachusetts Institute of Technology's Committee on Academic Responsibility included in Volume II of this report.

2. See, for example, Section 50.105 of the final PHS rule on responsibilities for dealing with possible misconduct (DHHS, 1989a, p. 32451):

> Institutions shall foster a research environment that discourages misconduct in all research and that deals forthrightly with possible misconduct associated with research for which PHS funds have been provided or requested. An institution's failure to comply with its assurance and the requirements of this

subpart may result in enforcement action against the institutions, including loss of funding, and may lead to the OSI's conducting its own investigation.

3. The policy, issued jointly by NIH and ADAMHA, became effective July 1, 1990. See National Institutes of Health and Alcohol, Drug Abuse, and Mental Health Administration (1989, 1990).

4. The deadline for the first applications affected by this rule was January 10, 1991. Thus at the time this report is being written, these applications are being reviewed. It will therefore be some months before the initial impact of the new requirement can be reviewed. For the NIH's initial thoughts on compliance, see Department of Health and Human Services (1990b).

5. See, for example, the experience in biomedical ethics reported in Culver et al. (1985).

6. Much of this section draws on a paper prepared for the panel by Nicholas Steneck, "Fostering Responsible Conduct in Science and Engineering Research: Current University Policies and Actions," which is included in Volume II of this report.

7. Other institutions have not adopted policies on integrity or responsibility, but they have adopted rules of academic discipline. See especially the compendium of student honor codes in *Codes and Regulations*, published as part of the Princeton Conference on Honor Systems, March 1988.

8. For more information on university policies and the research environment, see, in Volume II of this report, Barbara Mishkin's "Factors Enhancing Acceptance of Federal Regulation of Research" and Nicholas Steneck's "Fostering Responsible Conduct in Science and Engineering Research: Current University Policies and Actions." For examples, see policy statements from Harvard University School of Medicine, the University of Michigan, the Johns Hopkins University, and the University of California, San Diego, also in Volume II of this report.

9. See Department of Health and Human Services (1989d). Also see National Science Foundation (1990b) and prior semiannual reports (NSF, 1989c, 1990a).

10. See Harvard University Faculty of Medicine (1988), University of Michigan Medical School (1989), Johns Hopkins University School of Medicine (1990), and National Institutes of Health (1990).

11. Many researchers and academic administrators report a positive experience with other institutional policies that define appropriate research behavior. This is particularly true with the regulations for research involving human subjects and regulations on laboratory safety.

12. It is useful to review the findings presented in Institute of Medicine (1989a). The IOM report states:

> *Increasing budgetary and competitive pressures in science demand that local research institutions and government research funders develop standards to ensure responsible research practices to ensure the integrity of the academic research enterprise.* [emphasis in original]
>
> [The IOM committee expressed] consensus that, although the fundamental values and standards of the research community are appropriate, the expression and implementation of these standards are insufficient to promote responsible research practices in an increasingly large, heterogeneous, and competitive research environment. New and comprehensive guidelines should be developed by the research community to clarify traditional practices, to strengthen the mix of formal policies and informal practices currently in place, and to correct actions that seriously deviate from these standards.

13. The full texts of these institutional guidelines and additional examples are included in Volume II of this report.

7

Recommendations

Ensuring the integrity of the research process requires that scientists and research institutions give systematic attention to the fundamental values, principles, and traditions that foster responsible research conduct. In considering factors that may affect integrity and misconduct in science, the panel formulated the following twelve recommendations to strengthen the research enterprise and to clarify the nature of the responsibilities of scientists, research institutions, and government agencies in this area.

ACTING TO DEFINE AND STRENGTHEN BASIC PRINCIPLES AND PRACTICES

Recommendation One

Individual scientists and officials of research institutions should accept formal responsibility for ensuring the integrity of the research process. They should foster an environment, a reward system, and a training process that encourage responsible research practices.

Discussion: Scientists and research institutions need to accept formal responsibility for ensuring the integrity of the research process. Although faculty and research staff have the primary responsibility for maintaining integrity, institutional officials should retain

145

and accept certain explicit obligations. Institutions should strive to attain a research enterprise that emphasizes and rewards excellence in science, quality rather than quantity, openness rather than secrecy, and collegial obligations rather than opportunistic behavior in appointment, promotion, tenure, and other career decisions.

However, aggressive efforts to assure responsible research practices, if carried to an extreme, can damage the research enterprise. Balance is required. Inflexible rules or requirements can increase the time and effort necessary to conduct research, can decrease innovation, can discourage creative individuals from pursuing research careers, and can in some instances make the research process impossible.

In particular, mentors and research directors should (1) educate themselves, their students, and associates about responsible research practices; (2) examine difficult or problematic issues that provide opportunities to clarify principles, rights, interests, and obligations that may come into conflict; and (3) inform their students and associates about available institutional channels for expressing concerns regarding misconduct in science, questionable research practices, and other misconduct.

Efforts to improve the research training experience need encouragement. The research community should recognize the damage that can be done by poor mentorship practices, whether abusive or neglectful. Inappropriate practices should be identified and corrected quickly, but with regard for the privacy of the involved parties. Institutional leaders should take steps to establish a climate within the research setting that encourages research collaboration and educational training and fosters constructive ties between mentors and trainees. This climate should encourage the identification of poor mentorship practices at an early stage and establish fall-back arrangements in case some unanticipated event—such as the death of a mentor, or an instance of misconduct in science or other misconduct—disturbs the relationship. Fall-back provisions should provide necessary support, both emotional and material, to the trainee from the resources of the department or institution.

Recommendation Two

Scientists and research institutions should integrate into their curricula educational programs that foster faculty and student awareness of concerns related to the integrity of the research process.

Discussion: Educational programs on research ethics should reflect the diverse perspectives of the scientific community but should focus on identifying fundamental principles that guide responsible research practices. Educators and scientists should suggest how these principles can help resolve ethical dilemmas associated with specific research practices, provide information about relevant laws and regulations that govern misconduct in science and other misconduct in the research environment, and discuss the historical development of good scientific practice.

Recommendation Three

Adoption of formal guidelines for the conduct of research, which can provide a valuable opportunity for faculty and research institutions to clarify the nature of responsible research practices, should be an option, not a requirement, for research institutions.

Discussion: In principle, guidelines for the conduct of research should be framed to fit local situations, including specific research fields and protocols, and should be formulated by the scientists who conduct research, since they know the specific matters relevant to their work. Guidelines should be actively discussed by all who are affected by them and modified as experience dictates.

DEALING WITH MISCONDUCT— INSTITUTIONAL ROLES

Recommendation Four

Research institutions and government agencies should adopt a common framework of definitions for distinguishing among misconduct in science, questionable research practices, and other forms of misconduct. They should adopt a single consistent definition of misconduct in science that is based on fabrication, falsification, and plagiarism. Accordingly, the panel recommends that federal agencies review their definitions of misconduct in science to remove ambiguous categories such as "other serious deviations from accepted research practices."

Recommendation Five

Government agencies should adopt common policies and procedures for handling allegations of misconduct in science. The Office of Science and Technology Policy (OSTP) should lead the effort to establish government-wide definitions and procedures. OSTP

should consider adopting the definition of misconduct in science proposed in Chapter 1 of this report and use this definition in formulating government-wide model policies.

Discussion: The variation among existing regulatory definitions of misconduct in science is a serious obstacle to developing effective local institutional polices for handing allegations of misconduct in science. The ambiguity of the scope of regulatory definitions provides opportunity for serious misunderstandings between individual scientists and research institutions and between institutions and government agencies.

A federal interagency committee has been established by OSTP to establish model policies and procedures for government agencies to use in handling allegations of misconduct in science. That committee plans to recommend a basic definition of misconduct in science and basic procedures for developing uniform approaches to addressing misconduct-in-science cases. However, it is not certain that this recommendation will eliminate significant differences among governmental agency definitions of misconduct in science, especially differences between those adopted by the National Science Foundation and the Public Health Service.

The model policies and procedures under consideration by OSTP provide a significant opportunity to reduce inconsistencies, uncertainties, and conflicts in current governmental procedures.

Recommendation Six

Research institutions and government research agencies should have policies and procedures that ensure appropriate and prompt responses to allegations of misconduct in science. Research institutions should foster effective and appropriate methods for detecting and handling incidents of misconduct in science and should strengthen the implementation of misconduct-in-science policies and procedures that incorporate fundamental elements of due process.

Discussion: Research institutions that receive funding from the Public Health Service are already legally required to have policies and procedures to handle alleged and confirmed misconduct in science. Such policies and procedures should be extended to all public and private research conducted within these institutions and should incorporate the following essential requirements:

1. Clear communication to members of the research community about effective institutional channels for reporting misconduct in science without fear of retribution.

2. Resources and means for handling accusations and investigations with dispatch. Response times should be short within the limits of other responsibilities.

3. Fair and impartial treatment for all who are involved in cases. Both accused and accuser are entitled to anonymity during the early phases of misconduct-in-science investigations.

4. Protection of the legal rights of those involved in cases throughout the proceedings. Additional protections may be necessary for faculty members who participate as witnesses or members of investigative panels.

5. Provision for encouraging, through a variety of educational activities, open discussions of misconduct in science, questionable research practices, and other misconduct, so that all members of the research community know clearly what is expected of them.

Recommendation Seven

Scientists and their institutions should act to discourage questionable research practices through a broad range of formal and informal methods in the research environment. They should also accept responsibility for determining which questionable research practices are serious enough to warrant institutional penalties. But the methods used by individual scientists and research institutions to address questionable research practices should be distinct from those for handling misconduct in science or other misconduct.

Recommendation Eight

Research institutions should have policies and procedures to address other misconduct—such as theft, harassment, or vandalism—that may occur in the research environment. Where procedures for handling complaints about other misconduct do not exist, allegations should be examined according to the same administrative mechanisms as those designed to address misconduct in science, although the procedural pathways for responding to other misconduct and misconduct in science may differ.

Recommendation Nine

Government research agencies should clarify their roles in addressing misconduct in science, other misconduct, and questionable research practices. Although government agencies have specific regulatory responsibilities in handling the categories of misconduct in science and other misconduct, their role in addressing questionable research practices should be designed to support the efforts of scientists and research institutions to discourage such practices through the processes of education and peer review.

Discussion: The role of government in fostering research integrity should be restricted to one that supports and facilitates the efforts of scientists and their institutions. Efforts to standardize research practices across the disciplines or across institutions should be avoided, since they may weaken many of the strengths and positive features associated with the diversity that fosters intellectual freedom in the research environment.

TAKING ADDITIONAL STEPS

Recommendation Ten

An independent Scientific Integrity Advisory Board should be created by the scientific community and research institutions to exercise leadership in addressing ethical issues in research conduct; in framing model policies and procedures to address misconduct in science and other misconduct; to collect and analyze data on episodes of misconduct in the research environment; to provide periodic assessments of the adequacy of public and private systems that have been developed to handle misconduct-in-science cases; and to facilitate the exchange of information about and experience with policies and procedures governing the handling of allegations of misconduct in science.

Discussion and Details

The new private, not-for-profit organization recommended by the panel would *not* address specific cases of alleged misconduct in science, nor would it accredit procedures adopted by particular institutions. The purpose of the organization—called the Scientific Integrity Advisory Board (SIAB)—would be to provide an information resource and advice for performers and sponsors of research. SIAB would also increase public awareness of how the scientific research enterprise works and highlight areas of consensus and disagreement.

Specific Tasks

The Scientific Integrity Advisory Board would provide resources to accomplish the following:

1. Address and frame ethical issues arising in the conduct of research. Through consultations with research scientists and other concerned individuals and institutions, SIAB could identify issues associated with data handling, authorship, and training and mentorship, as well as conflicts between research faculty and institutional officials.

2. Formulate model policies and procedures for handling allegations of misconduct in science consistent with the standards of due process and confidentiality that should govern the handling of complaints. In fulfilling its charge, SIAB would comment on selected features of misconduct-in-science policies and procedures adopted by research institutions and, upon request, provide a confidential evaluation of proposed policies to the subject institution. SIAB could also monitor the activities of the Office of Science and Technology Policy, the Public Health Service, the National Science Foundation, and other agencies to review and comment publicly on agency policies and procedures for handling allegations of misconduct in science.

3. Provide advisory opinions and guidance to institutional officials regarding points of difficulty or uncertainty that might arise in implementing policies and procedures for handling misconduct in science. Examples of checkpoints to be addressed include the manner in which an individual accused of misconduct should be notified about the nature of the allegation; the status and protection of witnesses and supporting evidence; formulation of the charge to panels examining allegations of misconduct; criteria to be considered in appointing members of an investigating panel; notification of research sponsors, potential employers, and other individuals who might be affected by the outcome of an investigation of alleged misconduct; and the legal status of documents and individuals associated with misconduct inquiries and investigations.

4. Develop a set of case studies illustrating problem areas and inadequate safeguards in handling complaints involving misconduct in science or other misconduct in the research environment.

5. Provide guidance to institutions about appropriate actions to be taken in responding to allegations of misconduct in science or other misconduct.

6. Act as a clearinghouse for literature and records involving incidents of misconduct in science. This information could be derived from media accounts, court proceedings, government documents that are subject to requests under the Freedom of Information Act, and investigative reports that would be provided voluntarily by research institutions. SIAB could prepare a series of white papers or reports to summarize this information.

Organization and Structure

The panel believes that SIAB should be an independent board composed of practicing scientists, research administrators, individuals who have reported and handled incidents of misconduct, former government officials, and public figures who are not involved in the scientific enterprise. The board of SIAB should include individuals who are knowledgeable about one or more of the various scientific disciplines and should also be constructed to assure that the experiences of diverse institutions are taken into account in SIAB activities.

The governance structure should assure objectivity and independence, the critical ingredients for SIAB's success. Although it is important that SIAB maintain its credibility as an independent organization, it may be necessary to establish SIAB within an existing entity to provide institutional stability and to facilitate its interaction with a broad network of public and private officials. A suitable host organization should be considered to enable SIAB to develop the details of a charter, operating plan, and budget and to receive start-up funding. SIAB might or might not become part of the host organization once it began operation.

The Scientific Integrity Advisory Board could operate in the same manner as the Carnegie Commission on Science, Technology, and Government, founded in 1988 by the Carnegie Corporation of New York as a nongovernmental organization to assess the process by which the government incorporates scientific and technical knowledge into policy and decision making.[1] The commission includes former government officials, eminent scientists, and private sector leaders as well as an advisory council. It organizes studies, issues interim reports, makes final recommendations, and evaluates the impact of its work.

A possible host for SIAB is the National Academy of Public Administration (NAPA), which has a congressional charter. The elected membership of NAPA includes many distinguished scientists and former public officials with significant public service experience. Other organizations that have experience in handling medical or other professional malpractice cases may also be potential host organizations.

The National Academy of Sciences (NAS) can play an important role in facilitating the creation of SIAB, although it should not be viewed as a potential host. The NAS can provide a neutral forum to review and evaluate the ultimate purposes of and potential sponsors for SIAB. But the NAS does not have the resources or experience that would be necessary to provide the types of operational and advisory services that should be an integral part of SIAB's structure.

The host organization should be one that can ensure SIAB's standing and expertise while at the same time providing necessary independence from the research community that it would serve. This last criterion should specifically exclude as possible hosts professional societies that represent colleges and universities or the academic research community. SIAB's information services would be available to participating research universities and colleges, as well as to nonprofit hospitals and research laboratories. Private industry and government laboratories could be part of its audience as well, to the extent that their interests converged with those of academic research institutions. Initial funding might be provided by private foundations and interested federal agencies; operational funds should eventually come from diverse sources, including research institutions that subscribe to SIAB's information services or participate in SIAB-sponsored events.

The panel suggests that SIAB begin operation with an executive director, one full-time professional staff member, and one full-time support staff member. Additional staff members might be needed in the course of SIAB's providing active services and programs to research institutions and the public.

Termination

The panel believes that SIAB has the potential to improve significantly both the performance of the scientific community in handling instances of misconduct and also the environment for sustaining integrity in research. The panel strongly wishes to avoid creating another layer of bureaucracy and, accordingly, suggests that SIAB be authorized for an initial 5-year period only. It should automatically cease to exist unless an independent evaluation informs the directors of SIAB that continuation of its efforts is desired.

Concluding Comments

The proposal to establish SIAB deserves the support of research institutions, sponsoring federal agencies, and Congress. The research institutions should welcome the assistance of SIAB in establishing "best practice" policies and procedures. SIAB would not replace or interfere with the principal responsibility of research institutions to deal with specific cases of alleged scientific misconduct.

Agencies that sponsor research should welcome SIAB as a new and independent mechanism to ensure the continued integrity of the U.S. scientific research enterprise. These agencies should recognize the need to inform scientists whose research they sponsor as fully as possible of their procedures, the results of their investigations, and the nature of sanctions that are imposed.

Congress should view SIAB as a concrete step taken by the scientific community to establish a new and objective means to respond to concerns about the integrity of federally sponsored research. SIAB offers a mechanism that would foster informed judgments about how well *both* research performers and sponsors are progressing in ensuring the integrity of the research process.

Finally, the activities of SIAB could ensure that appropriate policies and procedures were being followed and thus could contribute to public confidence in the integrity of science and in the principles of self-governance according to which the scientific research enterprise operates.

Recommendation Eleven

The important role that individual scientists can play in disclosing incidents of misconduct in science should be acknowledged. Individuals who, in good conscience, report suspected misconduct in science deserve support and protection. Their efforts, as well as the efforts of those who participate in misconduct proceedings, can be invaluable in preserving the integrity of the research process. When necessary, serious and considered whistle-blowing is an act of courage that should be supported by the entire research community.

Discussion: All scientists have a responsibility to report suspected misconduct in science to appropriate authorities as part of their professional obligations. Just as it is essential to have procedural protections for individuals accused of misconduct in science, so also is it essential to protect individuals who report misconduct as well as those who participate in fact finding and adjudication to resolve allegations of misconduct. When a prolonged investigation is expected, research managers may suggest a temporary reassignment of both the subject of the investigation and the complainant during the time of the inquiry and investigation to mitigate possible tensions. Research institutions and, in some cases, scientific societies may also offer to provide professional recommendations for persons who have been instrumental in disclosing misconduct if the proceedings prove time consuming. But these supportive efforts should not be a replacement for the professional and personal support that is necessary from individual members of the scientific community. In particular, when serious breaches of professional standards and practices have occurred, it is important that senior scientists speak out publicly to validate and explain the legitimacy of the complaints.

Recommendation Twelve

Scientific societies and scientific journals should continue to provide and expand resources and forums to foster responsible research practices and to address misconduct in science and questionable research practices.

NOTE

1. For a discussion of the commission's origins, see Carnegie Commission on Science, Technology, and Government (1991).

8

Selected Bibliography

The following bibliography is a condensed version of more than 1,400 reference items in the panel's project files. This version omits references to news articles, book reviews, draft papers, and early government reports that have been replaced by revised policies.

Abelson, J., R. C. Atkinson, R. L. Davidson, R. Dulbecco, M. E. Friedkin, G. E. Palade, J. E. Seegmiller, M. I. Simon, and D. Steinberg. 1990. "PNAS paper: rules and standard practice." *Science* 249(September 21):1358.

Abelson, P. 1990. "Mechanisms for evaluating scientific information and the role of peer review." *Journal of the American Society for Information Science* 41(3):216-22.

Agricultural Research Service (ARS). 1990. *Procedures for Reporting and Dealing with Possible Misconduct in Science.* Draft report. ARS, U.S. Department of Agriculture, Beltsville, Md.

Allison, P. D. and J. S. Long. 1987. "Interuniversity mobility of academic scientists." *American Sociological Review* 52(October):643-52.

Altman, L. and L. Melcher. 1983. "Fraud in science." *British Medical Journal* 286(June 25):2003-06.

American Association for the Advancement of Science (AAAS). 1990. *AAAS Resolution on Federal Legislation Providing Immunity for Investigations and Reporting of Scientific Fraud and Misconduct,* approved by the Board of Directors, April 27. AAAS, Washington, D.C.

American Association for the Advancement of Science (AAAS). 1991a. *AAAS Report XVI: Research and Development FY1992.* AAAS, Washington, D.C.

American Association for the Advancement of Science (AAAS). 1991b. "Misconduct in science: Recurring issues, fresh perspectives." Executive summary of conference, November 15-16, Washington, D.C.

American Association for the Advancement of Science-American Bar Association (AAAS-ABA) National Conference of Lawyers and Scientists. 1989. *Project on Scientific Fraud and Misconduct*. Reports on workshops one (September 18-20, 1987), two (September 23-25, 1988), and three (February 17-18, 1989). Three volumes. AAAS, Washington, D.C.

American Association of University Professors (AAUP). 1989. "Statement on plagiarism." *Academe* (September/October):47-48.

American Bar Association (ABA). n.d. *A Survey on the Teaching of Professional Responsibility*. Center for Professional Responsibility, ABA, Chicago, Ill.

American Chemical Society (ACS). 1986. "Ethical guidelines to publication of chemical research." Pp. 217-22 in *The ACS Style Guide*, Dodd, J. S. (ed.). ACS, Washington, D.C.

American Chemical Society (ACS). 1989. *ACS Academic Professional Guidelines*. ACS, Washington, D.C.

American Federation for Clinical Research (AFCR). 1989. "AFCR guidelines for the responsible conduct of research." *Clinical Research* 37(September):510-11.

American Society of Mechanical Engineers (ASME). 1988. *Society Policy: Ethics*. P-15.7. ASME, New York.

American Sociological Association (ASA). 1989. *Code of Ethics*. ASA, Washington, D.C.

American Statistical Association (ASA), Ad Hoc Committee on Professional Ethics. 1983. "Ethical Guidelines for Statistical Practice." *The American Statistician* 37(February):5-6. [With comments.]

American Statistical Association (ASA). 1989. *Ethical Guidelines for Statistical Practice*, Committee on Professional Ethics. ASA, Washington, D.C.

Amundson, N. R. 1987. "American university graduate work." *Chemical Engineering Education* 21(4):160-63.

Andersen, Robert. 1988. "The federal government's role in regulating misconduct in scientific and technological research." *Journal of Law and Technology* 3:121-148.

Angell, M. 1983. "Editors and fraud." *CBE Views* 6(2):3-8.

Angell, M. 1986. "Publish or perish: a proposal." *Annals of Internal Medicine* 104(2):261-62.

Angell, M. and A. S. Relman. 1988. "Fraud in biomedical research: a time for congressional restraint." *New England Journal of Medicine* 318(June 2):1462-63.

Annals of Internal Medicine. 1982. "Authorship from the reader's side." *Annals of Internal Medicine* 97(October):613-14.

Appelbaum, P. S., L. H. Roth, C. W. Lidz, P. Benson, and W. Winslade. 1987. "False hopes and best data: consent to research and the therapeutic misconception." *Hastings Center Report* (April):20-24.

Arkin, H. R. 1985a. "Academic dismissals: due process, part 1." *Journal of the American Medical Association* 254(November 1):2463-66.

Arkin, H. R. 1985b. "Academic dismissals: due process, part 2." *Journal of the American Medical Association* 254(November 8):2653-56.

Association of Academic Health Centers (AAHC). 1990. *Conflicts of Interest in Academic Health Centers*, AAHC Task Force on Science Policy. Policy paper 1. AAHC, Washington, D.C.

Association of American Medical Colleges (AAMC). 1982. *The Maintenance of High Ethical Standards in Conduct of Research*, Ad Hoc Committee on the Maintenance of High Ethical Standards in the Conduct of Research. AAMC, Washington, D.C.

Association of American Medical Colleges (AAMC). 1990. *Guidelines for Dealing with Faculty Conflicts of Commitment and Conflicts of Interest in Research*. AAMC, Washington, D.C.

Association of American Universities (AAU). 1983. *Report of the Association of American Universities Committee on the Integrity of Research*, Association of American Universities, American Council on Education, and National Association of State Universities and Land-Grant Colleges. AAU, Washington, D.C.

Association of American Universities (AAU). 1988. *Indirect Costs Associated with Federal Support of Research on University Campuses: Some Suggestions for Change*, AAU Ad Hoc Committee on Indirect Costs. AAU, Washington, D.C.

Babbage, C. 1989. "The decline of science in England." *Nature* 340(August 17):499-502.

Bailar, J. C., III. 1985. "When research results are in conflict." *New England Journal of Medicine* 313(October 24):1080-81.

Bailar, J. C., III. 1986. "Science, statistics, and deception." *Annals of Internal Medicine* 104(February):259-60.

Bailar, J. C., III and K. Patterson. 1985. "Journal peer review: the need for a research agenda." *New England Journal of Medicine* 312(March 7):654-57.

Bailar, J. C., III, M. Angell, S. Boots, E. S. Myers, N. Palmer, M. Shipley, and P. Woolf (eds.). 1990. *Ethics and Policy in Scientific Publication*. Council of Biology Editors, Bethesda, Md.

Baird, L. L. "The melancholy of anatomy: the personal and professional development of graduate and professional school students." *Higher Education: Handbook of Theory and Research*. Agathon Press, New York, in press.

Baltimore, D. 1989. "Baltimore's travels." *Issues in Science and Technology* 5(Summer):48-54.

Barber, A. A. 1989. "Misconduct in science: an administrator's view." Council of Graduate Schools meeting, Banff, Alberta, March 5.

Barber, B. 1978. "Control and responsibility in the powerful professions." *Political Science Quarterly* 93(4):599-615.

Barber, B. and W. Hirsch (eds.). 1962. *The Sociology of Science*. The Free Press of Glencoe, New York.

Barzun, J. 1978. "The professions under seige: private practice versus public need." *Harper's* (October):61-67.

Batshaw, M. L., L. P. Plotnick, B. G. Petty, P. K. Woolf, and E. D. Mellits. 1988. "Academic promotion at a medical school." *New England Journal of Medicine* 318(March 24):741-46.

Bayles, M. D. 1981. *Professional Ethics*. Wadsworth, Belmont, Calif.

Beaty, H. N., D. Babbott, E. J. Higgins, P. Jolly, and G. S. Levey. 1986. "Research activities of faculty in academic departments of medicine." *Annals of Internal Medicine* 104(January):90-97.

Bechtel, H. K., Jr. and W. Pearson, Jr. 1985. "Deviant scientists and scientific deviance." *Deviant Behavior* 6:237-252.

Berardo, F. M. 1989. "Scientific norms and research publication issues and professional ethics." *Sociological Inquiry* 59(3):249-66.

Bernstein, F. C., T. F. Koetzle, J. B. Williams, E. F. Meyer, Jr., M. D. Brice, J. R. Rodgers, O. Kennard, T. Shimanouchi, and M. Tasumi. 1977. "The protein data bank." *European Journal of Biochemistry* 80:319-24.

Beveridge, W. I. B. 1957. *The Art of Scientific Investigation*. Third edition. Vintage Books, New York.

Biven, L. W. 1989. "Scientific integrity programs of the Public Health Service." *Assistance Management Journal* 5(Fall):21-28.

Blair, C. and W. Schaffer. 1991. "Promotion of the responsible conduct of research." *NIH Peer Review Notes* (June):4-6.

Blisset, M. 1972. *Politics in Science.* Basic Studies in Politics. Little, Brown and Company, Boston.

Bloch, E. 1989. "Changes in NSF proposal format." *IMS Bulletin* 18(6):577.

Bloch, E. 1991. "Academic research funding: is there a crisis? Optimists, pessimists and realists." AAAS Colloquium on Science and Technology Policy, Washington, D.C., April 11.

Bloom, F. 1989. "Who should police scientific misconduct? Most misconduct is already resolved within the system." *Journal of NIH Research* 1(May/June):14.

Bok, D. 1988. "Ethics, the university, and society." *Harvard Magazine* (May/June):39-50.

Bok, D. 1990. *Universities and the Future of America.* Duke University Press, Durham, N.C.

Bosk, C. L. 1979. *Forgive and Remember: Managing Medical Failure.* University of Chicago Press, Chicago.

Bowen, W. G. and Sosa, J. A. 1989. *Prospects for Faculty in the Arts and Sciences: A Study of Factors Affecting Demand and Supply, 1987 to 2012.* Princeton University Press, Princeton, N.J.

Bowen, W. G., G. Lord, and J. A. Sosa. 1991. "Measuring time to doctorate: reinterpretation of the evidence." *Proceedings of the National Academy of Sciences* 88(February 1):713-17.

Boyer, E. L. 1990. *The Condition of the Professoriate: Attitudes and Trends, 1989,* Carnegie Foundation for the Advancement of Teaching. Princeton University Press, Princeton, N.J.

Bradshaw, R. A., A. J. Adler, F. Chytil, N. D. Goldberg, and B. J. Litman. 1990. "Bridges committee procedures." *Science* 250(November 2):611.

Braunwald, E. 1987. "On analysing scientific fraud." *Nature* 325(January 15):215-16.

Braxton, J. M. 1992. "The influence of graduate department quality on the sanctioning of scientific misconduct." *Journal of Higher Education,* forthcoming.

Brett, J. M., S. B. Goldberg, and W. L. Ury. 1990. "Designing systems for resolving disputes in organizations." *American Psychologist* 45(February):162-70.

Bridgstock, M. 1982. "A sociological approach to fraud in science." *Australia-New Zealand Journal of Science* 18(3):364-83.

Brigham and Women's Hospital. 1989. *Policies on Hospital-Industry Relationships.* Brigham and Women's Hospital, Boston.

Broad, W. J. 1981. "Fraud and the structure of science." *Science* 212(April 10):137-41.

Broad, W. and Wade, N. 1982. *Betrayers of the Truth: Fraud and Deceit in the Halls of Science.* Simon and Schuster, New York.

Brooks, H. 1989. "Lessons of history: Successive challenges to science policy." In *The Research System in Transition,* Cozzens, S. E., P. Healey, A. Rip, and J. Ziman, (eds.). NATO Advanced Studies Institute Series, Series D, Volume 57. Kluwer Academic Publishers, Boston.

Browning, G. 1991. "Testing the hypothesis." *National Journal* (May 4):1078.

Bulger, R. E. 1988. "The need for an ethical code for teachers of the basic biomedical sciences." *Journal of Medical Education* 63(February):131-33.

Burks, C. and L. Tomlinson. 1989. "Submission of data to GenBank." *Proceedings of the National Academy of Sciences* 86(January):408.

Burman, K. D. 1982. "'Hanging from the masthead': reflections on authorship." *Annals of Internal Medicine* 97(4):602 05.

Burton, R. V. 1981. "Landmarks in the literature: Can ethics be taught?" *New York University Education Quarterly* 12(3):29-32.

Busch, J. W. 1985. "Mentoring in graduate schools of education: mentors' perceptions." *American Educational Research Journal* 22(2):257-65.

Bush, V. 1945. *Science, the Endless Frontier: A Report to the President.* Office of Scientific Research and Development. U.S. Government Printing Office, Washington, D.C.

Calabrese, R. L. and J. T. Cochran. 1990. "The relationship of alienation to cheating among a sample of American adolescents." *Journal of Research and Development in Education* 23(Winter):65-72.

Calkins, E. V., L. M. Arnold, and T. L. Willoughby. 1987. "Perceptions of the role of a faculty supervisor or 'mentor' at two medical schools." *Assessment and Evaluation in Higher Education* 12(3):202-08.

Calkins, E. V., L. M. Arnold, T. L. Willoughby, and S. C. Hamburger. 1986. "Docents' and students' perceptions of the ideal and actual role of the docent." *Journal of Medical Education* 61(September):743-48.

Callahan, D. and S. Bok. 1980. *Ethics Teaching in Higher Education.* Plenum Press, New York.

Carnegie Commission on Science, Technology, and Government. 1991. Science and Technology and the President. Carnegie Commission, New York.

Casper, G. 1990. "University policy documents." *The University of Chicago Chronicle* (October 25):2.

Cassidy, M. M. and A. E. Shamoo. 1989. "First international conference on scientific data audit, policies and quality assurance." *Accountability in Research: Policies and Quality Assurance* 1(1):1-3.

Ceci, S. J. 1988. "Scientists' attitudes toward data sharing." *Science, Technology, and Human Values* 13(1/2):45-52.

Ceci, S. J. and E. Walker. 1983. "Private archives and public needs." *American Psychologist* 38(4):414-23.

Cesa, I. L. and S. C. Fraser. 1989. "A method for encouraging the development of good mentor-protege relationships." *Teaching of Psychology* 16(3):125-28.

Chalk, R. 1978. "Scientific society involvement in whistleblowing." *Science, Technology, and Human Values* (January):47-51.

Chalk, R. and P. Woolf. 1989. "Regulating a 'knowledge business'." *Issues in Science and Technology* 5(Winter):33-35.

Chalk, R., M. Frankel, and S. Chafer. 1981. *Professional Ethics Activities in Science and Engineering Societies.* American Association for the Advancement of Science, Washington, D.C.

Charrow, R. 1989. "Who should police scientific misconduct? Each new scandal increases the chance for federal regulation." *Journal of NIH Research* 1(May/June):15.

Charrow, R. P. 1990. "Scientific misconduct revisited: OSI on trial." *Journal of NIH Research* 2(October):83-85.

Charrow, R. P. 1991. "Message to NIH's Office of Scientific Integrity: here comes the judge." *Journal of NIH Research* 3(February):97-99.

Chubin, D. E. 1985. "Misconduct in research: an issue of science policy and practice." *Minerva* 23(Summer):175-202.

Chubin, D. E. 1988. "Allocating credit and blame in science." *Science, Technology, and Human Values* 13(1/2):53-63.

Chubin, D. E. 1990. "Scientific malpractice and the contemporary politics of knowledge." Pp. 144-63 in *Theories of Science in Society*, Cozzens, S. E. and T. F. Gieryn (eds.). Indiana University Press, Bloomington, Ind.

Chubin, D. E. and E. J. Hackett. 1990. *Peerless Science: Peer Review and U.S. Science Policy.* SUNY Series in Science, Technology, and Society. SUNY Press, Albany, N.Y.

Cinkosky, M. J., J. W. Fickett, P. Gilna, and C. Burks. 1991. "Electronic data publishing and GenBank." *Science* 252(May 31):1273-77.

Cohen, D. L., L. B. McCullough, R. W. I. Kessel, A. Apostolides, K. Heiderich, and E. R. Alden. 1988. "A national survey concerning the ethical aspects of informed consent and role of medical students." *Journal of Medical Education* 63(November):821-29.

Cole, J. F. and J. A. Lipton. 1977. "The reputations of American medical schools." *Social Forces* 55:662-84.

Cole, S., L. Rubin, and J. R. Cole. 1977. "Peer review and the support of science." *Scientific American* 237(4):34-41.

Cordes, C. 1990. "U.S. enters lawsuit accusing scientist, institutions of fraud." *Chronicle of Higher Education* (September 5):A1.

Council for Advancement and Support of Education (CASE). 1988. *Special Advisory for College and University Presidents*, National Task Force on Higher Education and the Public Interest. CASE, Washington, D.C.

Council of Graduate Schools (CGS). 1990a. *Research Student and Supervisor: An Approach to Good Supervisory Practice.* CGS, Washington, D.C.

Council of Graduate Schools (CGS). 1990b. *Summary of Oral and Written Reports Presented at the Meeting on the Role and Nature of the Doctoral Dissertation*, January 26. CGS, Washington, D.C.

Council of Graduate Schools (CGS). 1990c. *The Doctor of Philosophy Degree: A Policy Statement*, Task Force on the Doctor of Philosophy Degree. CGS, Washington, D.C.

Council of Graduate Schools (CGS). 1991. *The Role and Nature of the Doctoral Dissertation: A Policy Statement.* CGS, Washington, D.C.

Cournand, A. 1977. "The code of the scientist and its relationship to ethics." *Science* 198(November 18):699-705.

Cournand, A. and M. Meyer. 1976. "The scientist's code." *Minerva* (Spring):79-96.

Cozzens, S. E. 1989. *Social Control and Multiple Discovery in Science: The Opiate Receptor Case.* SUNY Press, Albany, New York.

Cozzens, S. E., P. Healey, A. Rip, and J. Ziman (eds.). 1990. *The Research System in Transition.* NATO Advanced Studies Institute Series, Series D, Volume 57. Kluwer Academic Publishers, Boston.

Cram, D. 1989. "Commentary: Tribe and leader." *CGS Communicator* 22(April):1.

Crawford, S. and L. Stucki. 1990. "Peer review and the changing research record." *Journal of the American Society for Information Science* 41(April):223-28.

Cronan-Hillix, T., L. K. Gensheimer, W. A. Cronan-Hillix, and W. S. Davidson. 1986. "Students' views of mentors in psychology graduate training." *Teaching of Psychology* 13(3):123-27.

Crutcher, K. A. 1991. "How to succeed in science." *Perspectives in Biology and Medicine* 34(Winter):213-18.

Culliton, B. J. 1990. "NIH misconduct probes draw legal complaints." *Science* 249(July 20):240-43.

Culver, C. M., K. D. Clouser, B. Gert, H. Brody, J. Fletcher, A. Jonsen, L. Kopelman, J. Lynn, M. Siegler, and D. Wikler. 1985. "Special report: basic curricular goals in medical ethics." *New England Journal of Medicine* 312(January 24):253-56.

Dakins, D. R. 1989. "Copyright lawsuit illuminates debate over academic standards." *Diagnostic Imaging* (May):54-60.

Dana–Farber Cancer Institute. 1987. "Policy for recording and preserving scientific data." October. Dana-Farber Cancer Institute, Boston, Mass.

Daniels, G. H. 1967. "The pure-science ideal and democratic culture." *Science* 156(June 30):1699-1706.

David, R. J. and J. A. Kovach. 1979. "Attitudes towards unethical behavior as a function of educational commercialization." *College Student Journal* (Winter):338-44.

Davis, B. D. 1989. "Government and quality in science." *Science* 246(November 10):736.

Davis, M. S. 1989. *The Perceived Seriousness and Incidence of Ethical Misconduct in Academic Science.* Unpublished Ph.D. thesis. Ohio State University, Columbus.

Davis, R. D. 1988. "New censors in the academy: two approaches to curb their influence." *Science, Technology, and Human Values* 13(1&2):64-74.

Department of Health and Human Services (DHHS). 1985. *ALERT System for Institutions, Organizations and Individuals Under Investigation for Possible Misconduct in Science or Subject to Sanctions for Such Misconduct.* October 15 draft. Public Health Service, Washington, D.C.

Department of Health and Human Services (DHHS). 1986. "Policies and procedures for dealing with possible misconduct in science." *NIH Guide for Grants and Contracts* 15(July 18):1-37.

Department of Health and Human Services (DHHS). 1989a. "Responsibilities of PHS awardee and applicant institutions for dealing with and reporting possible misconduct in science: final rule." *Federal Register* 54(August 8):32446-32451.

Department of Health and Human Services (DHHS). 1989b. "Statement of organization, functions and delegations of authority." Public Health Service (PHS). *Federal Register* 54(March 16):11080-81.

Department of Health and Human Services (DHHS). 1989c. *Draft Standards for the Responsible Conduct of Science in PHS Intramural Research Programs.* Office of Scientific Integrity Review, Washington, D.C.

Department of Health and Human Services (DHHS). 1989d. *Misconduct in Scientific Research,* Office of Inspector General. OAI-88-07-00420. Washington, D.C.

Department of Health and Human Services (DHHS). 1990a. *Policies and Procedures for Dealing with Possible Misconduct in Extramural Research.* August 30. Public Health Service, Washington, D.C.

Department of Health and Human Services (DHHS). 1990b. *PHS Workshop: Education and Training of Scientists in the Responsible Conduct of Research.* March 8-9. Public Health Service, Washington, D.C.

Department of Health and Human Services (DHHS). 1990c. *Workshop on Data Management in Biomedical Research: Background Papers.* April 25-26. Public Health Service, Washington, D.C.

Department of Health and Human Services (DHHS). 1991a. "Proposed policies and procedures for dealing with possible scientific misconduct in extramural research." *Federal Register* 56(June 13):27384-94.

Department of Health and Human Services (DHHS). 1991b. *First Annual Report: Scientific Misconduct Investigations Reviewed by Office of Scientific Integrity Review, March 1989 - December 1990.* Office of Scientific Integrity Review, Washington, D.C.

Department of Health and Human Services (DHHS). 1991c. *Data Management in Biomedical Research: Report of a Workshop,* Office of Scientific Integrity Review, Public Health Service, Washington, D.C.

Department of Health and Human Services (DHHS). 1991d. Minutes, Meetings of the OSIR Advisory Committee. Office of Scientific Integrity Review. July 15 and November 17. OSIR, Washington, D.C.

Dewald, W. G., J. G. Thursby, and R. G. Anderson. 1986. "Replication in empirical economics: The Journal of Money, Credit and Banking project." *American Economic Review* 76(4):587-603.

Diamond, A. L. and D. R. Laurance. 1985. "Product liability in respect of drugs." *British Medical Journal* 290(February 2):365-68.

Dickinson, J. P. 1984. *Science and Scientific Researchers in Modern Society.* UNESCO, Paris.

Djerassi, C. 1986. "Castor's dilemma." *The Hudson Review* 39(3):405-18.

Djerassi, C. 1989. *Cantor's Dilemma.* Doubleday. New York.

Donaldson, V. H. 1984. "When things are not as they seem." *The Journal of Laboratory and Clinical Medicine* 103(April):491-96.

Dong, E. 1991. "Confronting scientific fraud." *The Chronicle of Higher Education* (October 9):A52.

Duke University. 1988. *Code of Policy and Procedures for Dealing with Allegations of Scientific Fraud or Misconduct.* Duke University, Durham, N.C.

Dworkin, G. 1983. "Fraud and science." Pp. 65-74 in *Research Ethics.* Alan R. Liss, Inc., New York. [Reprinted from *Progress in Clinical Biological Research,* Volume 128, 1983.]

Edsall, J. T. 1975. *Scientific Freedom and Responsibility,* Committee on Scientific Freedom and Responsibility. American Association for the Advancement of Science, Washington, D.C.

Edsall, J. T. 1981. "Two aspects of scientific responsibility." *Science* 212(April 3):11-14.

Eisenberg, R. S. 1989. "Patents and the progress of science: exclusive rights and experimental use." *University of Chicago Law Review* 56(Summer):1017-86.

Eisenhut, L. P. 1990. "Universität prüft Anschuldigungen gegen Professorin." *Kolner Stadtanzeiger* (October 24).

Ellenberg, J. H. 1983. "Ethical guidelines for statistical practice: a historical perspective." *The American Statistician* 37(February):1-4.

Emory University. 1991. *Draft Proposal for a University-wide Policy for Research Conduct: Guidelines for the Conduct of Research and Scholarship,* March 20 draft. Emory University, Atlanta, Ga.

Engler, R. L., J. W. Covell, P. J. Friedman, P. S. Kitcher, and R. M. Peters. 1987. "Misrepresentation and responsibility in medical research." *New England Journal of Medicine* 317(November 26):1383-89.

Epstein, R. 1986. "On drafting rules and procedures for academic fraud." *Minerva* 24(Summer-Autumn):344-46.

Erkut, S. and J. R. Mokros. 1984. "Professors as models and mentors for college students." *American Educational Research Journal* 21(2):399-417.

Eser, A. 1987. "Researcher as "offender" and "victim"—comparative observations as to freedom and responsibility of science and technology." *Brigham Young University Law Review* 1987(2):571-85.

Etnier, C. 1986. "Secrecy and the young researcher." *Technology in Society* 8:267-71.

Evanoski, P. O. 1988. "The role of mentoring in higher education." *Community Review* 8(2):22-27.

Fass, R. A. 1990. "Cheating and plagiarism." Pp. 170-84 in *Ethics and Higher Education*, May, W. M. (ed.). American Council on Education. Macmillan, New York.

Feinstein, R. J. 1985. "The ethics of professional regulation." *New England Journal of Medicine* 312(March 21):801-04.

Feynman, Richard. 1985. *Surely You're Joking, Mr. Feynman.* W. W. Norton, New York.

Fletcher, R. 1987. "The doubtful case of Cyril Burt." *Social Policy & Administration* 21(1):40-57.

Florman, S. C. 1978. "Moral blueprints: on regulating the ethics of engineers." *Harper's* (October):30-33.

Foelsing, A. 1984. *Der Mogelfaktor.* Hamburg, pp. 20-21.

Fox, R. 1985. "Reflections and opportunities in the sociology of medicine." *Journal of Health and Social Behavior* 26(March):6-14.

Francis, J. R. 1990. "The credibility and legitimation of science: a loss of faith in the scientific narrative." *Accountability in Research: Policies and Quality Assurance* 1(1):5-22.

Frankel, M. 1988. "In the news." *Professional Ethics Report* (Fall):1.

Frankel, M. S. 1989. "Professional codes: why, how, and with what impact?" *Journal of Business Ethics* 8:109-15.

Frazer, M. J. and A. Kornhauser (eds.). 1986. *Ethics and Social Responsibility of Science Education.* ICSU Press, New York.

Freedman, D. X. 1988. "The meaning of full disclosure." *Archives of General Psychiatry* 45(July):689-91.

Freidson, E. 1970. *Profession of Medicine: A Study of the Sociology of Applied Knowledge.* Dodd, Mead, and Company, New York.

Friedman, P. J. 1988. "Research ethics, due process, and common sense." *Journal of the American Medical Association* 260(October 7):1937-38.

Friedman, R. S. and R. C. Friedman. 1982. *Role of University Organized Research Units in Academic Science.* NSF/PRM 82-004. National Science Foundation, Washington, D.C.

Fruton, J. S. 1990. *Contrasts in Scientific Style: Research Groups in the Chemical and Biochemical Sciences.* American Philosophical Society, Philadelphia.

Galtung, J. 1983. "Researchers, elites, and people in a rapidly changing world." Pp. 95-108 in *Research Ethics.* Alan R. Liss, Inc., New York.

Gaston, J. 1978. "Disputes and deviant views about the ethos of science." In *The Reward System in British and American Science.* John Wiley and Sons, New York.

Geiger, R. L. 1990. "The American University and Research." Pp. 15-35 in *The Academic Research Enterprise within the Industrialized Nations: Comparative Perspectives*, Government-University-Industry Research Roundtable. National Academy Press, Washington, D.C.

Geis, G. and R. F. Meier (eds.). 1977. *White-collar Crime: Offenses in Business, Politics, and the Professions.* The Free Press, New York.

Gert, B., W. A. Nelson, and C. M. Culver. 1989. "Moral theory and neurology." *Neurologic Clinics* 7(November):681-96.

Gjerde, C. L. and S. E. Colombo. 1982. "Promotion criteria: perceptions of faculty members and departmental chairmen." *Journal of Medical Education* 57(March):157-62.

Glazer, M. P. and Glazer, P. M. 1990. *The Whistleblowers: Exposing Corruption in Government and Industry.* Basic Books, New York.

Glazer, S. 1988. "Combating science fraud." *CQ Editorial Research Reports* 2(August 5):390.

Glick, J. L. 1990. "On the potential cost effectiveness of scientific audits." *Accountability in Research: Policies and Quality Assurance* 1(1):77-83.

Gluck, M. E., D. Blumenthal, and M. A. Stoto. 1987. "University-industry relationships in the life sciences: implications for students and post-doctoral fellows." *Research Policy* 16:327-36.

Golberg, L. 1982. "A code of ethics for scientists reporting and reviewing information on chemicals." *Fundamental and Applied Toxicology* 2(11-12):289-92.

Goldschmidt, R. B. 1956. *Portraits from Memory: Recollections of a Zoologist.* University of Washington Press, Seattle.

Goodstein, D. 1991. "Scientific fraud." *American Scholar* 60(4):505-15.

Government–University–Industry Research Roundtable (GUIRR). 1989. *Science and Technology in the Academic Enterprise: Status, Trends, and Issues.* National Academy Press, Washington, D.C.

Green, H. P. 1987. "Scientific responsibility and the law." *Journal of Law Reform* 20(Summer):1009-27.

Greene, P. J., J. S. Durch, W. Horwitz, and V. Hooper. 1985. "Policies for responding to allegations of fraud in research." *Minerva* 23(Summer):203-215.

Grinnel, F. 1987. *The Scientific Attitude.* Westview, Boulder, Colo.

Group for the Advancement of Psychiatry (GAP). 1990. *A Casebook in Psychiatric Ethics*, GAP Committee on Medical Education. Brunner/Mazel, New York.

Grouse, L. D. 1982. "Dealing with alleged fraud in medical research." *Journal of the American Medical Association* 248(October 1):1637-38.

Haas, W. 1991. "An overview of optical disk imaging." Unpublished paper.

Hackett, E. J. 1990. "Science as a vocation in the 1990s: the changing organizational culture of academic science." *Journal of Higher Education* 61(3):241-79.

Hackett, E. J. and D. E. Chubin. 1990. "Scientific malpractice and the politics of knowledge." Annual meeting of the Society for Social Studies of Science, Minneapolis, October.

Hagstrom, W. O. 1965. *The Scientific Community.* Basic Books, New York.

Hall, E. C., C. M. Huguley, Jr., P. N. Symbas, and N. C. Moran. 1985. "Research of Dr. John R. Darsee at Emory University." *Minerva* 23(Summer):276-304.

Hallum, J. V. and S. W. Hadley. 1990. "OSI: why, what, and how." *ASM News* (12):647-51.

Hallum, J. V. and S. W. Hadley. 1991. "Editorial: rights to due process in instances of possible scientific misconduct." *Endocrinology* 128(2):643-44.

Hamilton, D. P. 1990. "PNAS bars papers from UC geneticist." *Science* 249(August 10):622.

Hamilton, D. P. 1991a. "Can OSI withstand a scientific backlash?" *Science* 253:1084-1086.

Hamilton, D. P. 1991b. "NIH finds fraud in Cell paper." *Science* 251(March 29):1552-54.

Hanna, K. E. 1989. "Collaborative research in biomedicine: resolving conflicts." Background paper, Institute of Medicine (IOM) Committee on Government-Industry Collaboration in Biomedical Research and Education, Washington, D.C.

Hansen, B. C. and K. D. Hansen. 1991a. "The proper role of the Office of Scientific Integrity: institutional vs. federal responsibilities." *FASEB Journal* 5(11):2507-2508.

Hansen, B. C. and K. D. Hansen. 1991b. "Scientific fraud and the Public Health Service Act: a critical analysis." *FASEB Journal* 5(11):2512-2515.

Hardy, R. J. and D. Burch. 1981. "What political science professors should know in dealing with academic misconduct." *Teaching Political Science* 9(Fall):5-14.

Harvard University Faculty of Medicine. 1988. *Guidelines for Investigators in Scientific Research*, Harvard University, Cambridge, Mass. [Also published in *Clinical Research* 37(April 1989):192-93.]

Harvard University Faculty of Medicine. 1990. *Policy on Conflicts of Interest and Commitment*. Harvard University, Cambridge, Mass., March 22.

Healy, B., L. Campeau, R. Gray, J. A. Herd, B. Hoogwerf, D. Hunninghake, G. Knatterud, W. Stewart, and C. White, et al. 1989. "Conflict-of-interest guidelines for a multicenter clinical trial of treatment after coronary-artery bypass-graft surgery." *New England Journal of Medicine* 320(April 6):949-51.

Heinz, L. and D. Chubin. 1988. "Congress investigates scientific fraud." *BioScience* (September):559-61.

Hendrickson, R. M. 1988. "Removing tenured faculty for cause." *Educational Law Reporter* 44(April 14):483-94.

Hill, S., E. Kogler, M. H. Bahniuk, and J. Dobos. 1989. "The impact of mentoring and collegial support on faculty success: an analysis of support behavior, information adequacy, and communication apprehension." *Communication Education* 38(January):15-33.

Hinder, S. and P. Halfpenny. 1990. "Potential for sharing research equipment." *Science and Public Policy* 17(2):105-10.

Hollis, B. W. 1987. "I turned in my mentor." *The Scientist* 1(December 14):11-12.

Holton, G. 1975. "On the role of themata in scientific thought." *Science* 188(April 25):328-34.

Holton, G. 1978. *The Scientific Imagination: Case Studies*. Cambridge University Press, New York.

Holton, G. 1988. *Thematic Origins of Scientific Thought: Kepler to Einstein*. Revised edition. Harvard University Press, Cambridge, Mass.

Holton, G. 1990. "The drive to unity in physics, and the drive to delegitimate science." April 17. Andrew Gemant Award Lecture at the American Physical Society, Washington, D.C.

Hoshiko, Tom. 1991. "Facing ethical dilemmas: Scientists must lead the charge." *The Scientist* (October 28):11, 13-14.

Hull, David. 1988. *Science as a Process: An Evolutionary Account of the Social and Conceptual Development of Science*. University of Chicago Press, Chicago.

Hunter, H. O. 1985. "Academic self-government in the United States." *Minerva* 23(Spring):1-28.

Huth, E. 1986a. "Irresponsible authorship and wasteful publication." *Annals of Internal Medicine* 104(2):258-65.

Huth, E. J. 1986b. "Abuses and uses of authorship." *Annals of Internal Medicine* 104(February):266-67.

Huth, E. J. 1986c. "Guidelines on authorship of medical papers." *Annals of Internal Medicine* 104(February):269-74.

Huth, E. J. 1988. *Scientific Authorship and Publication: Process, Standards, Problems, Suggestions*. Background paper, Institute of Medicine (IOM) Committee on the Responsible Conduct of Research, Washington, D.C.

Industrial Chemist. 1987a. "Do you ever fake a research result?" February.

Industrial Chemist. 1987b. "Error and fraud in the lab." May, p. 84.

Institute of Medicine (IOM). 1989a. *The Responsible Conduct of Research in the Health Sciences*, Committee on the Responsible Conduct of Research, Division of Health Sciences Policy. National Academy Press, Washington, D.C.

Institute of Medicine (IOM). 1989b. *Report of a Workshop*, Committee on Government-Industry Collaboration in Biomedical Research and Education. National Academy Press, Washington, D.C.

International Committee of Medical Journal Editors. 1985. "Style matters: guidelines on authorship." *British Medical Journal* 291(September 14):722.

International Committee of Medical Journal Editors. 1988. "Uniform requirements for manuscripts submitted to biomedical journals." *Annals of Internal Medicine* 108(February):258-66.

International Statistical Institute. 1986. "ISI declaration on professional ethics." *International Statistical Review* 54(2):227-42.

Jackson, C. I. and J. W. Prados. 1983. "Honor in science." *American Scientist* 71(Sept.-Oct.):462-64.

Jacobstein, J. 1987. "I am not optimistic." *The Scientist* 1(December 14):11-12.

Jasanoff, S. 1988. *Good Laboratory Practices: Regulating Responsibility in Science.* Background paper, Institute of Medicine (IOM) Committee on the Responsible Conduct of Research, Washington, D.C.

Jayaraman, K. S. 1991. "Gupta faces suspension." *Nature* 349(February 21):645.

Johns Hopkins University School of Medicine. 1989. *Policy on Conflict of Commitment and Conflict of Interest.* Johns Hopkins University, Baltimore, Md.

Johns Hopkins University School of Medicine. 1990. *Rules and Guidelines for Responsible Conduct of Research.* Johns Hopkins University, Baltimore, Md.

Jonsen, A. R. 1984. "Public policy and human research." Pp. 3-20 in *Biomedical Ethics Reviews,* Humber, J. M. and R. T. Almeder (eds.). Humana, Clifton, N.J.

Joynson, R. B. 1989. *The Burt Affair.* Routledge, London.

Kaplin, W. A. 1986. *The Law and Higher Education.* Jossey-Bass, San Francisco.

Karlins, M., C. Michaels, and S. Podlogar. 1988. "An empirical investigation of actual cheating in a large sample of undergraduates." *Research in Higher Education* 29(4):359-364.

Kass, L. R. 1983. "Professing ethically." *Journal of the American Medical Association* 249(March 11):1305-10.

Keenan, C. E., G. C. Brown, H. N. Pontell, and G. Geis. 1985. "Medical students' attitudes on physician fraud and abuse in the Medicare and Medicaid programs." *Journal of Medical Education* 60(March):167-73.

Kennedy, D. 1989. "Statement on academic authorship." *Stanford University Research Policy Handbook.* Stanford University, Stanford, Calif.

Kennedy, D. 1991. *The Improvement of Teaching.* Stanford University, Stanford, Calif.

Kleppner, D. 1991. "The mismeasure of science." *The Sciences* (May/June):18-21.

Klotz, I. M. 1985. *Diamond Dealers and Feather Merchants: Tales from the Sciences.* Birkhauser, Boston.

Knight, E. J. 1990. *Scientific Misconduct and Self-Deception.* Senior honors thesis. University of Colorado, Boulder, Colo.

Knight, J. 1991. "Scientific misconduct: the rights of the accused." *Issues in Science and Technology* 8(1):28-29.

Knight, J. A. 1984. "Exploring the compromise of ethical principles in science." *Perspectives in Biology and Medicine* 27(3):432-42.

Knorr-Cetina, K. D. and M. Mulkay (eds.). 1983. *Science Observed: Perspectives on the Social Study of Science.* Sage, Beverly Hills, Calif.

Knox, P. L. and T. V. McGovern. 1988. "Mentoring women in academia." *Teaching of Psychology* 15(1):39-41.

Knox, R. A. 1983a. "The Harvard fraud case: where does the problem lie?" *Journal of the American Medical Association* 249(April 8):1797-1806.

Knox, R. A. 1983b. "Deeper problems for Darsee: Emory probe." *Journal of the American Medical Association* 249(June 3):2867-76.

Kohn, A. 1986. *False Prophets: Fraud and Error in Science and Medicine.* Basil Blackwell, New York.

Koj, A. 1990. "Complex responsibilities of scientists in the contemporary world." *Science and Public Policy* 17(2):78-81.

Koshland, D. E. 1987. "Fraud in science." *Science* 235(January 9):141.

Koshland, D. E. 1988a. "Science, journalism, and whistle-blowing." *Science* 240(April 29):585.

Koshland, D. E. 1988b. "The price of progress." *Science* 241(August 5):637.

Koshland, D. E. 1990. "Conflict of interest." *Science* 249(July 13):109.

Kreytbosch, C. E. and S. L. Messinger. 1968. "Unequal peers: the situation of researchers at Berkeley." *American Behavioral Scientist* 11(5):33-43.

Kubie, L. S. 1953. "Some unsolved problems of the scientific career, part II." *American Scientist* (November):101-12.

Kuyper, B. J. 1991. "Bringing up scientists in the art of critiquing research." *Bioscience* (April):248-50.

Lane, J., R. Ray, and D. Glennon. 1990. "Work profiles of research statisticians." *American Statistician* 44(February):9-13.

Langmuir, I. "Pathological science: scientific studies based on non-existent phenomena." *Speculations in Science and Technology* 8(2):77-94.

Lanza-Kaduce, L. 1980. "Deviance among professionals: the case of unnecessary surgery." *Deviant Behavior* 1:333-59.

Laor, N. 1985. "Prometheus the impostor." *British Medical Journal* 290(March 2):681-84.

LaPidus, J. B. and B. Mishkin. 1990. "Values and ethics in the graduate education of scientists." Pp. 283-298 in *Ethics and Higher Education*, May, W. W. (ed.). American Council on Education. Macmillan, New York.

Latour, B. 1987. *Science in Action: How to Follow Scientists and Engineers Through Society.* Harvard University Press, Cambridge, Mass.

Latour, B. and S. Woolgar. 1979. *Laboratory Life: The Social Construction of Scientific Facts.* Sage Library of Social Research, 80. Sage, Beverly Hills, Calif.

Lauscher, S. 1987. "Scientific misconduct: a case study." *Grants Magazine* 10(3):143-49.

Lawson, A. J. 1989. "The organic chemist and the personal computer." *Impact of Science on Society* 39(4):291-302.

Lederman, L. M. 1990. *Science: The End of the Frontier?* Report to the Board of Directors of the American Association for the Advancement of Science, AAAS, Washington, D.C.

Lee, B. G. 1978. "Fraud and abuse in medicare and medicaid." *Administrative Law Review* 30:301-43.

Levy, C. S. 1979. "Professional ethics: dilemmas of code construction." National Conference of Social Welfare, Philadelphia, May 15.

Lewin, B. 1987. "Fraud in science: the burden of proof." *Cell* 48(January 16):1-2.

Lewin, B. 1989. "Fraud and the fabric of science." *Cell* 57(June 2):699-700.

Lind, S. E. 1986. "Fee-for-service research." *New England Journal of Medicine* 314(January 30):312-15.

Lipschutz, S. B. 1990. "Toward independent scholarship: collaborative research relationships in doctoral education." *CGS Communicator* (April):1-2.

List, C. J. 1985. "Scientific fraud: social deviance or the failure of virtue?" *Science, Technology, and Human Values* 10(4):27-36.

Littlejohn, M. J. and C. M. Matthews. 1989. *Scientific Misconduct in Academia: Efforts to Address the Issue.* 89-392 SPR. Congressional Research Service, Washington, D.C., June 30.

Lo, B. 1987. "Behind closed doors: promises and pitfalls of ethics committees." *New England Journal of Medicine* 317(July 2):46-50.

Lock, S. 1985. *A Difficult Balance: Editorial Peer Review in Medicine.* ISI Press, Philadelphia.

Lock, S. 1988a. "Fraud in medicine." *British Medical Journal* 296(February 6):376-77.

Lock, S. 1988b. "Scientific misconduct." *British Medical Journal* 297(September 24):1531-1535.

Lock, S. 1990. "Medical misconduct: A survey in Britain." In *Ethics and Policy in Scientific Publication*, Bailar et al. (eds.). Council of Biology Editors, Bethesda, Md.

Lock, S. 1991. "Where do we go from here?" *Nature* 350(April 11):530.

Long, J. S. and R. McGinnis. 1981. "Organizational context and scientific productivity." *American Sociological Review* 46(August):422-42.

Long, J. S. and R. McGinnis. 1985. "The effects of the mentor on the academic career." *Scientometrics* 7(3-6):255-280.

Long, J. S., P. D. Allison, and R. McGinnis. 1979. "Entrance into the academic career." *American Sociological Review* 44(5):816-30.

Lovain, T. B. 1983-1984. "Grounds for dismissing tenured postsecondary faculty for cause." *Journal of College and University Law* 10(3):419-33.

Lundberg, G. D. and A. Flanagin. 1989. "New requirements for authors: signed statements of authorship responsibility and financial disclosure." *Journal of the American Medical Association* 262(October 13):2003-05.

Maddox, J. 1988a. "Why the pressure to publish?" *Nature* 333(June 9):493.

Maddox, J. 1988b. "Can a Greek tragedy be avoided?" *Nature* 333(June 30):795-97.

Maddox, J. 1989a. "Making good databanks better." *Nature* 341(September 28):277.

Maddox, J. 1989b. "Is the salami sliced too thin?" *Nature* 342(December 14):733.

Maddox, J. 1989c. "Making authors toe the line." *Nature* 342(December 21/28):855.

Maddox, J. 1990a. "What to do with extraneous data." *Nature* 346(July 19):215.

Maddox, J. 1990b. "Should camp-followers be policemen?" *Nature* 348(November 8):107.

Maddox, J. 1991. "Secret Service as ultimate referee." *Nature* 350(April 18):553.

Majerus, P. W. 1982. "Fraud in medical research." *Journal of Clinical Investigation* 70(July):213-17.

Manwell, C. and C. M. Ann Baker. 1981. "Honesty in science: a partial test of a sociobiological model of the social structure of science." *Search* 12(6):151-59.

Martin, B. 1986. "Bias in awarding research grants." *British Medical Journal* 293(August 30):550-52.

Martin, H. C., R. M. Ohmann, and J. Wheatly. 1969. *The Logic and Rhetoric of Exposition.* Third edition. Holt, Rinehart, and Winston, New York.

Martin, R. G. 1990. "Scientific hype." *New Biologist* 2(9):747-49.

May, W. F. 1980. "Professional ethics: setting, terrain, and teacher." Pp. 205-41 in *Ethics Teaching in Higher Education*, Callahan, D. and S. Bok (eds.). Plenum Press, New York.

May, W. W. (ed.). 1990. *Ethics and Higher Education.* American Council on Education. Macmillan, New York.

Mayr, Ernst. 1982. *The Growth of Biological Thought.* Harvard University Press, Cambridge, Mass.

Mayr, Ernst. 1988. *Toward a New Philosophy of Biology.* Harvard University Press, Cambridge, Mass.

Mazlish, B. 1982. "The quality of 'The Quality in Science': An evaluation." Pp. 48-67 in *Quality in Science*, La Follette, M. C. (ed.). MIT Press, Cambridge, Mass.

Mazur, Allan. 1989. "Allegations of dishonesty in research and their treatment by American universities." *Minerva* 27:177-194.

McBride, G. 1974. "The Sloan-Kettering affair: could it have happened anywhere?" *Journal of the American Medical Association* 229(September 9):1391-1410.

McCain, K. W. 1991. "Communication, competition and secrecy." *Science, Technology, and Human Values* 16(4):491-516.

McGarity, T. O. 1990. "Public participation in data audits." *Accountability in Research: Policies and Quality Assurance* 1(1):47-52.

McGinnis, R. and J. S. Long. 1988. "Entry into academia: effects of stratification, geography, and ecology." Pp. 28-51 in *Academic Labor Markets and Academic Careers in American Higher Education*, Breneman, D. W. and T. I. K. Youn (eds.). Falmer, New York.

McGinnis, R., P. D. Allison, and J. S. Long. 1982. "Postdoctoral training in bioscience: allocation and outcome." *Social Forces* 60(3):701-22.

Meadow, C. T. 1990. "Perspectives on . . . Evaluation of scientific information: peer review and the impact of new information technology, introduction and overview." *Journal of the American Society for Information Science* 41(April):214-15.

Medawar, P. 1964. "Is the scientific paper fraudulent? yes; it misrepresents scientific thought." *Saturday Review* (August 1):42-43.

Medawar, P. 1984. *The Limits of Science*. Harper and Row, New York.

Medawar, P. B. 1979. *Advice to a Young Scientist*. Alfred P. Sloan Foundation Series. Harper and Row, New York.

Menninger, D. C. 1990. "Scientific misconduct and the federal response." Paper presented to the annual meeting of the American Political Science Association, San Francisco, September 2.

Merriam, S. B., T. K. Thomas, and C. P. Zeph. 1987. "Mentoring in higher education: what we know." *The Review of Higher Education* 11(2):199-210.

Merton, R. K. 1973. *The Sociology of Science: Theoretical and Empirical Investigations*, Storer, N. W. (ed.). University of Chicago Press, Chicago.

Miceli, M. P., J. B. Dozier, and J. P. Near. 1991. "Blowing the whistle on data fudging: a controlled field experiment." *Journal of Applied Social Psychology* 21(4):271-95.

Miers, M. L. 1985. "Current NIH perspectives on misconduct in science." *American Psychologist* 40(7):831-35.

Mishkin, B. 1988. "Responding to scientific misconduct: due process and prevention." *Journal of the American Medical Association* 260(October 7):1932-36.

Mishkin, B. 1991a. "First annual report of scientific misconduct investigations reviewed by OSIR—a review and analysis." *Professional Ethics Report* (Spring):305.

Mishkin, B. 1991b. "Cooperation between academia and industry: opportunities and challenges." Fidia Research Foundation Ethical Issues in Research, Washington, D.C., March 29-30.

Mitroff, I. I. 1974. *The Subjective Side of Science: A Philosophical Inquiry into the Psychology of the Apollo Moon Scientists*. Elsevier, Amsterdam.

Mulkay, M. J. 1975. "Norms and ideology in science." *Social Science Information* 15(4/5):637-56.

Mulkay, M. and G. N. Gilbert. 1982. "Accounting for error: how scientists construct their social world when they account for correct and incorrect belief." *Sociology* 16:165-72.

National Academy of Sciences (NAS). 1989. *On Being A Scientist*, Committee on the Conduct of Science. National Academy Press, Washington, D.C.

National Academy of Sciences and National Research Council (NAS-NRC). 1984. *Science and Creationism: A View from the National Academy of Sciences.* National Academy Press, Washington, D.C.

National Center for Education Statistics (NCES). 1990. *Faculty in Higher Education Institutions, 1988.* NCES 90-365. Office of Education Research and Improvement, U.S. Department of Education. U.S. Government Printing Office, Washington, D.C.

National Commission for the Protection of Human Subjects of Biomedical and Behavioral Research. 1978. *Report and Recommendations: Institutional Review Boards.* DHEW (OS) 78-0008. U.S. Government Printing Office, Washington, D.C.

National Institutes of Health (NIH). 1987. "NIH policy relating to reporting and distribution of unique biological materials produced with NIH funding." *NIH Guide for Grants and Contracts* 16(October 23):1-3.

National Institutes of Health (NIH). 1988a. "Misconduct in science assurance." *NIH Guide for Grants and Contracts* 17(January 15):1-2.

National Institutes of Health (NIH). 1988b. "PHS policy relating to distribution of unique research resources produced with PHS funding." *NIH Guide for Grants and Contracts* 17(September 16):1-2.

National Institutes of Health. 1988c. *Retreat Summary Report: Retreat on Conflict of Interest in Collaborations with Industry.* December 19-20. Office of Invention Development, Bethesda, Md.

National Institutes of Health (NIH). 1990. *Guidelines for the Conduct of Research at the National Institutes of Health.* NIH, Bethesda, Md.

National Institutes of Health (NIH). 1991. *NIH Data Book.* NIH 91-126. September. Bethesda, Md.

National Institutes of Health (NIH) and Alcohol, Drug Abuse, and Mental Health Administration (ADAMHA). 1989. "Requirement for programs on the responsible conduct of research in national research service award institutional training programs." *NIH Guide for Grants and Contracts* 18(December 22):1.

National Institutes of Health (NIH) and Alcohol, Drug Abuse, and Mental Health Administration (ADAMHA). 1990. "Reminder and update: requirement for programs on the responsible conduct of research in national research service award institutional training programs." *NIH Guide for Grants and Contracts* 19(August 17):1.

National Research Council (NRC). 1981. *Postdoctoral Appointments and Disappointments,* Committee on a Study of Postdoctorals in Science and Engineering in the United States. National Academy Press, Washington, D.C.

National Research Council (NRC). 1985. *Sharing Research Data,* Fienberg, S. E., M. E. Martin, and M. L. Straf (eds.). National Academy Press, Washington, D.C.

National Research Council (NRC). *Summary Report.* 1989. *Doctorate Recipients from United States Universities,* Office of Science and Engineering Policy. National Academy Press, Washington, D.C.

National Science Board (NSB). 1988. *Report of the NSB Committee on Openness of Scientific Communication.* National Science Foundation, Washington, D.C.

National Science Board (NSB). 1989. *Science and Engineering Indicators.* NSB 89-1. National Science Foundation, Washington, D.C.

National Science Foundation (NSF). 1987. "Misconduct in science and engineering research: final regulations." *Federal Register* 52(July 1).24466-24470.

National Science Foundation (NSF). 1989a. *Report of NSF Staff Committee on Sharing.* Office of the General Counsel, NSF, Washington, D.C.

National Science Foundation (NSF). 1989b. "Important Notice to Presidents of Colleges and Universities and Heads of Other National Science Foundation Grantees Organizations." Notice No. 106. April 17. NSF, Washington, D.C.

National Science Foundation (NSF). 1989c. *Semiannual Report to the Congress.* Number 1. Office of Inspector General, NSF, Washington, D.C.

National Science Foundation (NSF). 1990a. *Semiannual Report to the Congress.* Number 2. Office of Inspector General, NSF, Washington, D.C.

National Science Foundation (NSF). 1990b. *Semiannual Report to the Congress.* Number 3. Office of Inspector General, NSF, Washington, D.C.

National Science Foundation (NSF). 1990c. *The State of Academic Science and Engineering.* NSF, Washington, D.C.

National Science Foundation (NSF). 1990d. *Survey Data on the Extent of Misconduct in Science and Engineering.* OIG-90-3214. Office of Inspector General, NSF, Washington, D.C.

National Science Foundation (NSF). 1991a. *Semiannual Report to the Congress.* Number 4. Office of Inspector General, NSF, Washington, D.C.

National Science Foundation (NSF). 1991b. "Misconduct in science and engineering: final rule." *Federal Register* 56(May 14):22286-90.

National Science Foundation (NSF). 1991c. *Semiannual Report to the Congress.* Number 5. Office of Inspector General, NSF, Washington, D.C.

Neidle, E. A. 1985. "The mentor apprentice program: a modest proposal for alleviating the scarcity of clinical research in dentistry." *Journal of Dental Education* 49(5):272-74.

Nelkin, D. 1983. "Whistle blowing and social responsibility in science." Pp. 351-57 in *Research Ethics.* Alan R. Liss, Inc., New York. [Reprinted from *Progress in Clinical Biological Research*, Volume 128, 1983.]

Nelkin, D. 1984. *Science as Intellectual Property.* Macmillan, New York.

New England Journal of Medicine. 1992. "Information for authors." April 2.

Newhouse, R. C. 1982. "Alienation and cheating behavior in the school environment." *Psychology in the Schools* 19(April):234-37.

Nobel, J. J. 1990. "Comparison of research quality guidelines in academic and nonacademic environments." *Journal of the American Medical Association* 263(March 9):1435-37.

Novack, J. M. and A. C. Bennett. 1983. "Informally inviting moral development: teacher perception and behaviour regarding the handling of moral transgressions." Annual Meeting of the American Educational Research Association, Montreal.

Nowack, J. A. 1989. *The University-Policy Environment for University-Industry Interactions.* Background paper, Committee on Government-Industry Collaboration in Biomedical Research and Education, Institute of Medicine, Washington, D.C.

Nowak, T., R. Rotunda, and J. Young. 1983. *Constitutional Law.* Second edition. West Publishing, St. Paul, Minn.

Nuss, E. M. "Academic integrity: comparing faculty and student attitudes." *Improving College and University Teaching* 32(3):140-44.

Oak Ridge Associated Universities (ORAU). 1990. *Code of Ethics and Standards of Conduct.* ORAU 90/J-125. Department of Energy, ORAU, Oak Ridge, Tenn.

Office of Technology Assessment (OTA). 1986a. *The Regulatory Environment for Science.* Technical Memorandum. OTA, Washington, D.C.

Office of Technology Assessment (OTA). 1986b. *Research Funding as an Investment: Can We Measure the Returns?* OTA-TM-SET-36. U.S. Government Printing Office, Washington, D.C.

Office of Technology Assessment (OTA). 1990. "Proposal pressure in the 1980s: An indicator of stress on the federal research system." Staff paper. OTA, Washington, D.C., April.

Office of Technology Assessment (OTA). 1991. *Federally Funded Research: Decisions for a Decade.* OTA-SET-490. U.S. Government Printing Office, Washington, D.C..

O'Reilly, J. T. 1990. "More gold and more fleece: improving the legal sanctions against medical research fraud." *Administrative Law Review* 42(Summer):393-422.

Osmond, D. H. 1983. "Malice's wonderland: research funding and peer review." *Journal of Neurobiology* 14(2):95-112.

O'Toole, M. 1991. "Margot O'Toole's record of events." *Nature* 351(May 16):180-82.

Parmley, W. W. 1990. "21st Bethesda conference: ethics in cardiovascular medicine." *Journal of the American College of Cardiology* 16(July)(1):1-36.

Petersdorf, R. G. 1982. "Preventing and investigating fraud in research." *Journal of Medical Education* 57:880-81.

Petersdorf, R. G. 1986. "The pathogenesis of fraud in medical science." *Annals of Internal Medicine* 104(February):252-54.

Petersdorf, R. G. 1989. "A matter of integrity." *Academic Medicine* (March):119-23.

Phillip, Kathryn. 1991. "Making the transition from bench scientist to lab leader." *The Scientist* (April 1):18-19.

Pigman, Ward and Emmett B. Carmichael. 1950. "An ethical code for scientists." *Science* (June 16):643-647.

Pinch, T. 1990. "The role of scientific communities in the development of science." *Impact of Science on Society* (159):219-25.

Pincus, K. V. 1990. "Financial auditing and fraud detection: implications for scientific data audit." *Accountability in Research: Policies and Quality Assurance* 1(1):53-70.

Polanyi, M. 1951. *The Logic of Liberty.* International Library of Sociology and Social Reconstruction. Routledge and Kegan Paul, London.

Polanyi, M. 1962a. *Personal Knowledge: Towards a Post-critical Philosophy.* University of Chicago Press, Chicago.

Polanyi, M. 1962b. "The republic of science: its political and economic theory." *Minerva* 1(Autumn):54-73.

Pontell, H. N., P. D. Jesilow, and G. Geis. 1982. "Policing physicians: practitioner fraud and abuse in a government medical program." *Social Problems* 30(1):117-25.

Pratt, C. B. and G. W. McLaughlin. 1989. "An analysis of predictors of college students' ethical inclinations." *Research in Higher Education* 30(2):195-219.

Price, D. K. 1979. "The ethical principles of scientific institutions." *Science, Technology, and Human Values* 4(1):46-60. [Originally presented at Nobel Symposium No. 44, Stockholm, 1978.]

Price, D. K. 1985. *America's Unwritten Constitution: Science, Religion, and Political Responsibility.* Harvard University Press, Cambridge, Mass.

Princeton Conference on Honor Systems. 1988. *Codes and Regulations.* Princeton University, Princeton, N.J., February.

Rall, J. E. 1990. "Conflict of interest as it relates to intramural scientists at the National Institutes of Health." Presented to the Endocrine Society, Atlanta, June 20.

Raub, W. F. 1990. "Statement of the Acting Director, NIH." U.S. Congress, House of Representatives, Committee on Energy and Commerce, Subcommittee on Oversight and Investigations, Washington, D.C., April 30.

Raub, W. F. 1991. "Testimony." U.S. Congress, House of Representatives, Committee on Energy and Commerce, Subcommittee on Oversight and Investigations, Washington, D.C., March 6.

Ravetz, J. 1971. *Scientific Knowledge and Its Social Problems.* Clarendon Press, Oxford.

Rawles, B. A. 1980. *The Influence of a Mentor on the Level of Self-Actualization of American Scientists.* Unpublished Ph.D. thesis, Ohio State University, Columbus.

Relman, A. S. 1983. "Lessons from the Darsee affair." *New England Journal of Medicine* 308(June 9):1415-17.

Relman, A. S. 1984. "Responsibilities of authorship: where does the buck stop?" *New England Journal of Medicine* 310(April 19):1048-49.

Relman, A. S. 1985. "Professional regulation and the state medical boards." *New England Journal of Medicine* 312(March 21):784-85.

Relman, A. S. 1989. "Economic incentives in clinical investigation." *New England Journal of Medicine* 320(April 6):933-34.

Relman, A. S. 1990. "New 'information for authors'—and readers." *New England Journal of Medicine* 323(July 5):56, 72.

Remington, J. A. 1988. "Beyond big science in America: the binding of inquiry." *Social Studies of Science* 18:45-72.

Rennie, D. 1986. "Guarding the guardians: a conference on editorial peer review." *Journal of the American Medical Association* 256(November 7):2391-92.

Rennie, D. 1989. "Editors and auditors." *Journal of the American Medical Association* 261(May 5):2543-45. [With letters.]

Rest, J. R. 1988. "Can ethics be taught in professional schools? The psychological research." *Easier Said Than Done* (Winter):22-26.

Rider, R. E. 1991. "Saving the records of big science." *American Libraries* (February):166-68.

Rip, A. 1988. "Contextual transformations in contemporary science." Pp. 59-85 in *Keeping Science Straight: A Critical Look at the Assessment of Science and Technology,* Jamison, A. (ed.). Department of Theory of Science and Center for Interdisciplinary Studies, University of Gothenburg, Gothenburg, Sweden.

Rodman, H. 1970. "The moral responsibility of journal editors and referees." *The American Sociologist* 5(November):351-57.

Rodwin, M. A. 1989. "Physicians' conflicts of interest." *New England Journal of Medicine* 321(November 16):1405-08.

Roe, R. 1990. *A Bill: To Promote the Integrity of Scientific Research, and for Other Purposes.* U.S. Congress, House of Representatives. 101st Cong., 2nd sess., April 20. [Discussion draft; not introduced.]

Rosner, J. L. 1990. "Reflections of science as a product." *Nature* 345(May 10):108.

Ross, M., I. J. Winograd, and J. L. Cook. 1989. "Report to the Director [of the U.S. Geological Survey]." January 25. U.S. Geological Survey, Washington, D.C.

Rostand, J. 1960. *Error and Deception in Science: Essays on Biological Aspects of Life.* Translated from the French by A. J. Pomerans. Basic Books, New York.

Roth, J. A. 1966. "Hired hand research." *The American Sociologist* (August):190-96.

Rothschild, L. 1972. "Forty-five varieties of research (and development)." *Nature* 239(October 13):373-78.

Royal College of Physicians. 1991. *Fraud and Misconduct in Medical Research: Causes, Investigation and Prevention.* Royal College of Physicians of London, London.

Royal Society of Chemistry (RSC). 1988. *Guidance on Professional Conduct.* Burlington House, Royal Society of Chemistry, London.

Ruegg, W. 1986. "The academic ethos." *Minerva* 24(Winter):393-412.

Rutherford, D. G. and S. G. Olswang. 1981. "Academic misconduct: the due process rights of students." *National Association of Student Personnel Administrators* 19(Fall):12-16.

Sapp, J. 1990. *Where the Truth Lies: Franz Moewus and the Origins of Molecular Biology.* Cambridge University Press, Cambridge.

Scheers, N. J. and C. M. Dayton. 1987. "Improved estimation of academic cheating behavior using the randomized response technique." *Research in Higher Education* 26(1):61-69.

Schein, E. H. 1990. "Organizational culture." *American Psychologist* 45(February):109-119.

Schmaus, W. 1983a. "Fraud and negligence in science." *Connecticut Medicine* 47(March):155-58.

Schmaus, W. 1983b. "Fraud and the norms of science." *Science, Technology, and Human Values* 8(4):12-22.

Shader, R. I. and D. J. Greenblatt. 1987. "Authorship and coauthorship—working out the meaning." *Journal of Clinical Psychopharmacology* 7(5):293.

Shapiro, M. F. and R. P. Charrow. 1985. "Scientific misconduct in investigational drug trials." *New England Journal of Medicine* 312(March 14):731-36.

Shapiro, M. F. and R. P. Charrow. 1989. "The role of data audits in detecting scientific misconduct." *Journal of the American Medical Association* 261(17):2505-11.

Shils, E. 1990a. "The university world turned upside down: does confidentiality of assessment by peers guarantee the quality of academic appointment?" *Minerva* 28(3):324-34.

Shils, E. 1990b. "The university world turned upside down: does confidentiality of assessment by peers guarantee the quality of academic appointment? part II." *Minerva* 28(4):469-85.

Sigma Xi. 1986. *Honor in Science.* Second edition. Sigma Xi, New Haven, Conn.

Sigma Xi. 1987. *A New Agenda for Science.* Sigma Xi, New Haven, Conn.

Sigma Xi. 1989. *Sketches of the American Scientist.* Sigma Xi, New Haven, Conn.

Sindermann, C. J. 1985. *The Joy of Science: Excellence and its Rewards.* Plenum Press, New York.

Sindermann, C. J. 1987. *Survival Strategies for New Scientists.* Plenum Press, New York.

Smith, B. L. R. 1990. *American Science Policy Since World War II.* The Brookings Institution, Washington, D.C.

Smith, J. M., P. N. Chase, and J. J. Byrd. 1986. "A formalized mentor system in an educational setting." *Engineering Education* 76(4):216-18.

Snyder, R. G. 1988. *The Federal Research Assistantship: Indicators of Utilization in Graduate Research Training in Science and Engineering.* Council of Graduate Schools, Washington, D.C.

Soskolne, C. L. 1989. "Epidemiology: questions of science, ethics, morality, and law." *American Journal of Epidemiology* 129(January):1-18.

Sprague, R. L. 1987. "I trusted the research system." *The Scientist* 1(December 14):11-12.

Steelman, J. R. 1947. *Science and Public Policy: Administration of Research, a Report to the President,* President's Scientific Research Board. U.S. Government Printing Office, Washington, D.C.

Sterling, T. D. and J. J. Weinkam. 1990. "Sharing scientific data." *Communications of the ACM* 33(8):112-19.

Stewart, W. and N. Feder. 1987. "The integrity of the scientific literature." *Nature* 325(January 15):207-214.

St. James-Roberts, I. 1976a. "Are researchers trustworthy?" *New Scientist* 72(September 2):481-83.

St. James-Roberts, I. 1976b. "Cheating in science." *New Scientist* 72(November 25):466-69.

Storer, N. 1966. *The Social System of Science.* Holt, Rinehart and Winston, New York.

Stossel, T. P. 1985. "Reviewer status and review quality: experience of the Journal of Clinical Investigation." *New England Journal of Medicine* 312(March 7):658-59.

Stromholm, S. 1990. "Hero or villain? Prometheus reconsidered." *Science and Public Policy* 17(April):69-72.

Summerlin Peer Review Committee. 1974. *Report*. Memorial Sloan-Kettering Cancer Center, New York, May 17.

Sun, M. 1989. "Peer review comes under peer review." *Science* 244(May 26):910-12.

Sundstrom, E., K. P. De Meuse, and D. Futrell. 1990. "Work teams: applications and effectiveness." *American Psychologist* 45(2):120-33.

Sutton, J. R. 1984. "Organizational autonomy and professional norms in science: a case study of the Lawrence Livermore Laboratory." *Social Studies of Science* 14:197-224.

Swan, N. 1989. "Preventing and dealing with scientific fraud in Australia." *Medical Journal of Australia* 150(February 20):169-70.

Swazey, J. P. and S. R. Scher (eds.). 1981. *Whistleblowing in Biomedical Research: Policies and Procedures for Responding to Reports of Misconduct*. President's Commission for the Study of Ethical Problems in Medicine and Biomedical and Behavioral Research. U.S. Government Printing Office, Washington, D.C.

Swazey, J. P., K. S. Louis, and M. S. Anderson. 1989. "University policies and ethical issues in research and graduate education: highlights of the CGS deans' survey." *CGS Communicator* 22(March):1-3, 7-8.

Szilagyi, D. E. 1984. "The elusive target: truth in scientific reporting." *Journal of Vascular Surgery* (March):243-53.

Tangney, J. P. 1987. "Fraud will out—or will it?" *New Scientist* (August 6):62-63.

Thompson, J. J., B. S. Dreben, E. Holtzman, and R. B. Kreiser. 1983. "Corporate funding of academic research." *Academe* (November-December):18a-23a.

Tranoy, K. E. 1983. "Is there a universal research ethic?" Pp. 3-12 in *Research Ethics*. Alan R. Liss, Inc., New York.

Traweek, S. 1988. *Beamtimes and Lifetimes: The World of High Energy Physicists*. Harvard University Press, Cambridge, Mass.

Tribe, L. H. 1988. *American Constitutional Law*. Second edition. Foundation Press, Mineola, N.Y.

Tuckman, H., S. Coyle, and Y. Bae. 1990. *On Time to the Doctorate: A Study of the Lengthening Time to Completion for Doctorates in Science and Engineering*, Office of Scientific and Engineering Personnel. National Academy Press, Washington, D.C.

Turner, S. P. 1990. "Forms of patronage." Pp. 185-211 in *Theories of Science in Society*, Cozzens, S.E. and T. F. Gieryn (eds.). Indiana University, Bloomington, Ind.

University of Chicago. 1986. "Report of the Committee on Academic Fraud." *Minerva* 24(Summer-Autumn):347-358.

University of Maryland at Baltimore. 1989. Policies and Procedures Related to Allegations or Other Evidence of Academic Misconduct. University of Maryland, Baltimore, December 22.

University of Michigan Medical School. 1989. *Guidelines for the Responsible Conduct of Research*, Medical School Committee to Develop Guidelines for the Responsible Conduct of Research. University of Michigan, Ann Arbor.

U.S. Congress. 1981a. *Fraud in Biomedical Research*. House of Representatives, Committee on Science and Technology, Subcommittee on Investigation and Oversight. 97th Cong., 1st sess., March 31 and April 1. U.S. Government Printing Office, Washington, D.C.

U.S. Congress. 1981b. *National Cancer Institute Contracting and Procurement Procedures, 1981*. Senate, Committee on Labor and Human Resources. 97th Cong., 1st sess., June 2. U.S. Government Printing Office, Washington, D.C.

U.S. Congress. 1986. *Science and the Regulatory Environment.* House of Representatives, Committee on Science and Technology, Task Force on Science Policy. 99th Cong., 2nd sess., February 26 and March 5-6. U.S. Government Printing Office, Washington, D.C.

U.S. Congress. 1988a. *Scientific Fraud and Misconduct and the Federal Response.* House of Representatives, Committee on Governmental Operations, Subcommittee on Human Resources and Intergovernmental Relations. 100th Cong., 2nd sess., April 11. U.S. Government Printing Office, Washington, D.C.

U.S. Congress. 1988b. *Scientific Fraud and Misconduct in the National Institutes of Health Biomedical Grant Programs.* Hearings. House of Representatives, Committee on Energy and Commerce. April 12.

U.S. Congress. 1988c. *Federal Response to Misconduct in Science: Are Conflicts of Interest Hazardous to Our Health?* House of Representatives, Committee on Government Operations, Subcommittee on Human Resources and Intergovernmental Relations. 100th Cong., 2nd sess., September 29. U.S. Government Printing Office, Washington, D.C.

U.S. Congress. 1989a. *Fraud in NIH Grant Programs.* House of Representatives, Committee on Energy and Commerce, Subcommittee on Oversight and Investigations. 100th Cong., 2nd sess., April 12. Serial No. 100-189. U.S. Government Printing Office, Washington, D.C.

U.S. Congress. 1989b. *Scientific Fraud.* House of Representatives, Committee on Energy and Commerce, Subcommittee on Oversight and Investigations. 101st Cong., 1st sess., May 4, 9. Serial No. 101-64. U.S. Government Printing Office, Washington, D.C.

U.S. Congress. 1989c. *Is Science for Sale? Conflicts of Interest vs. the Public Interest.* House of Representatives, Committee on Governmental Operations, Subcommittee on Human Resources and Intergovernmental Relations. 101st Cong., 1st sess., June 13. Serial No. 20-969. U.S. Government Printing Office, Washington, D.C.

U.S. Congress. 1990a. *Are Scientific Misconduct and Conflicts of Interest Hazardous to Our Health?* H. Rpt. 101-688. House of Representatives, Committee on Government Operations. 101st Cong., 2nd sess. U.S. Government Printing Office, Washington, D.C.

U.S. Congress. 1990b. *Maintaining the Integrity of Scientific Research.* House of Representatives, Committee on Science, Space, and Technology, Subcommittee on Investigations and Oversight. 101st Cong., 1st sess., June 28. No. 73. U.S. Government Printing Office, Washington, D.C.

U.S. Congress. 1990c. *Scientific Fraud, Part 2.* House of Representatives, Committee on Energy and Commerce, Subcommittee on Oversight and Investigations. 101st Cong., 1st sess., April 30 and May 14. U.S. Government Printing Office, Washington, D.C.

Van Alstyne, W. 1970. "The constitutional rights of teachers and professors." *Duke Law Journal* (5):841-79.

Vanderpool, H. Y. and G. B. Weiss. 1987. "False data and the therapeutic misconception: two urgent problems in reseach ethics." *Hastings Center Report* (April):16-19.

Warwick, D. P. and T. F. Pettigrew. 1983. "Toward ethical guidelines for policy research." *Hastings Center Report* (February):9-16.

Washington University Senate Council. 1990. *Research Integrity Policy for Washington University.* Washington University, St. Louis, April 17.

Weber, M. 1946 [1918]. "Science as a vocation." Pp. 129-56 in *From Max Weber: Essays in sociology,* Gerth, H. and C. W. Mills (eds.). Oxford University Press, New York.

Weil, V. 1988. "Policy incentives and constraints on scientific and technical information." *Science, Technology, and Human Values* 13(1&2):17-26.

Weil, V. and J. W. Snapper. 1989. *Owning Scientific and Technical Information*. Rutgers University Press, New Brunswick, N.J.

Weinstein, D. 1979. "Fraud in science." *Social Science Quarterly* 59(4):639-52.

Weinstein, D. 1981. "Scientific fraud and scientific ethics." *Connecticut Medicine* 45(October):655-58.

Weiss, T. 1991a. "NIH grants research integrity amendments of 1991." *Congressional Record* (April 17):E1288-89.

Weiss, T. 1991b. "Too many scientists who 'blow the whistle' end up losing their jobs and careers." *Chronicle of Higher Education* (June 26):A36.

Westin, A. F. 1981. *Whistle-blowing!* McGraw-Hill, New York.

Wheeler, D. 1991. "U.S. has barred grants to 6 scientists in past 2 years." *Chronicle of Higher Education* (July 3).

Williams, J. S. Forthcoming. "The importance of preserving scientific data." In *Archival Management of Research Resources in the Health Field*, McCall, N. and L. A. Mix (eds.). Johns Hopkins University Press, Baltimore, Md.

Wilshire, B. 1990. "Professionalism as purification ritual." *Journal of Higher Education* 61(May/June 1990):280-93.

Wilson, E. Bright. 1952. *An Introduction to Scientific Research*. McGraw-Hill, New York.

Wilson, J. T. 1983. *Academic Science, Higher Education, and the Federal Government, 1950-1983*. University of Chicago Press, Chicago.

Wolins, L. 1962. "Responsibility for raw data." *American Psychologist* 17:657-58.

Woolf, P. K. 1981. "Fraud in science: how much, how serious?" *Hastings Center Report* 11(October):9-14

Woolf, P. K. 1986. "Pressure to publish and fraud in science." *Annals of Internal Medicine* 104(2):254-56.

Woolf, P. K. 1988a. "Deception in scientific research." *Jurimetrics Journal* 29(Fall):67-95.

Woolf, P. K. 1988b. "Science needs vigilance not vigilantes." *Journal of the American Medical Association* 260(October 7):1939-40.

Yamamoto, K. R. 1982. "Faculty members as corporate officers: does cost outweigh benefit?" Pp. 195-201 in *From Genetic Engineering to Biotechnology—The Critical Transition*, Whelan, W. J. and S. Black (eds.). John Wiley & Sons, New York.

Yankauer, A. 1989. "Editor's annual report—manuscript requirements." *American Journal of Public Health* 79(April):413-14.

Yolles, B. J., J. C. Connors, and S. Grufferman. 1986. "Sounding Board: Obtaining access to data from government-sponsored medical research." *New England Journal of Medicine* 315(December 25):1669-72.

Zastrow, C. H. 1970. "Cheating among college graduate students." *Journal of Educational Research* 64(December):157-60.

Zen, E-an. 1988. "Abuse of coauthorship: its implications for young scientists and the role of journals." *Geology* 16:292.

Ziman, J. 1978. *Reliable Knowledge: An Exploration of the Grounds for Belief in Science*. Cambridge University Press, New York.

Ziman, J. 1990. "Research as a career." Pp. 345-59 in *The Research System in Transition*, Cozzens, S. E., P. Healey, A. Rip, and J. Ziman (eds.). NATO Advanced Studies Institute Series, Series D, Volume 57. Kluwer Academic Publishers, Boston.

Zuckerman, H. 1968. "Patterns of name ordering among authors of scientific papers: a study of social symbolism and its ambiguity." *American Journal of Sociology* 74(November 3):276-291.

Zuckerman, Harriet. 1977. "Deviant behavior and social control in science." Pp. 87-138 in *Deviance and Social Change*. Sage, Beverly Hills, Calif.

Zuckerman, H. 1984. "Norms and deviant behavior in science." *Science, Technology, and Human Values* 9(1):7-13.

Zuckerman, H. 1988a. "The sociology of science." Pp. 511-73 in *Handbook of Sociology*, Smelser, N. J. (ed.). Sage, Newbury Park, Calif.

Zuckerman, H. A. 1988b. "Introduction: Intellectual property and diverse rights of ownership in science." *Science, Technology, and Human Values* 13(1&2):7-16.

Zuckerman, H. and R. K. Merton. 1971. "Patterns of evaluation in science: institutionalism, structure and functions of the referee system." *Minerva* 9(1):66-100.

Minority Statement

Three general concerns preclude our support of the present report. First, its overall tone presents an unbalanced treatment of scientists and institutions. It fails to convey the overriding importance of intellectual freedom and trust in a creative process that has been remarkably successful, and it lacks conviction in assessing the consequences of inappropriate institutional action or inaction. Second, the report is equivocal in defining misconduct in science and is inadequate in stating explicitly the problems inherent in alternative definitions. The "other misconduct" category introduces ambiguities into the definition, and blurs the boundaries between misconduct in science and questionable practice. Misconduct in science requires rigorous adjudicatory machinery and governmental oversight, protection of whistle-blowers, due process, strong sanctions, and full disclosure. In contrast, questionable practices raise issues about the value system and culture of science, and underscore the need for explicit dialogue and education. Governmental intervention is inappropriate for concerns regarding errors in collecting and interpreting data, incompetence, sloppiness, selection of data, authorship practices, multiple publications, and the like. The absence of consensus on the definition overtly undermines a primary goal of the report to achieve clear boundaries for the definition of misconduct in science. Third, the report does not stress sufficiently the importance of establishing a regularized institutional "response pathway" for allegations of misconduct, and for considering problems stemming from institutional and individual conflict of interest. Problematic institutional responses are a common theme in complicated cases of misconduct in science; yet the crucial need for and intricate complexities of vigorous, prompt and fair responses to allegations, establishing an "open door" from bench to Bethesda, are not emphasized. The report is weak in condemning the ALERT system of the Public Health Service which lists individuals because they are the subject of an investigation even though they should be presumed innocent. Finally, conflicts of interest directly related to research can be more complex, potentially more serious and perhaps more numerous than the examples of fabrication, falsification, and plagiarism, and therefore need to be addressed in this report.

Howard K. Schachman
Keith R. Yamamoto

December 30, 1991

APPENDIXES

A

Biographical Sketches of Panel Members

EDWARD E. DAVID, JR., D.Sc., the panel's chairman, is president of EED, Inc., in Bedminster, New Jersey. Educated as an electrical engineer, Dr. David previously served as White House Science Adviser (1970-1973) and was formerly president of Exxon Research and Engineering Company and research director of Bell Telephone Laboratories.

PHILIP H. ABELSON, Ph.D., is deputy editor, *Science*, and science advisor for the American Association for the Advancement of Science (AAAS). Dr. Abelson was the editor of *Science* for more than 20 years and is a recipient of the Presidential Medal of Science.

VICTOR R. BAKER, Ph.D., is Regents Professor and Professor of Geosciences and Planetary Sciences at the University of Arizona.

ALBERT BARBER, Ph.D., is vice chancellor for research at the University of California, Los Angeles.

MICHAEL BERMAN, J.D., is president of The Duberstein Group, Inc., Washington, D.C. He is an attorney and has extensive legislative experience. Mr. Berman was deputy chief of staff to Vice President Walter F. Mondale.

JOHN DEUTCH, Ph.D., is former provost and Institute Professor of Chemistry at the Massachusetts Institute of Technology. Dr. Deutch was formerly undersecretary of the Department of Energy.

VAL L. FITCH, Ph.D., is James S. McDonnell Distinguished University Professor of Physics, Joseph Henry Laboratories, Princeton University. He was awarded the Nobel Prize in physics in 1980.

MARYE ANNE FOX, Ph.D., is the M. June and J. Virgil Waggoner Regents Chair in Chemistry, University of Texas at Austin. She is an associate editor of the *Journal of the American Chemical Society* and a member of the National Science Board.

PETER GALISON, Ph.D., is co-chairman of the History of Science Program at Stanford University.

BERNARD GERT, Ph.D., is Stone Professor of Intellectual and Moral Philosophy at Dartmouth College.

IRA J. HIRSH, Ph.D., is Mallinckrodt Distinguished University Professor of Psychology and Audiology at Washington University in St. Louis, where he was formerly dean of the Faculty of Arts and Sciences. He was also formerly director of research at Central Institute for the Deaf.

JENNY L. McFARLAND, Ph.D., is a postdoctoral fellow in the Department of Brain and Cognitive Sciences at the Massachusetts Institute of Technology.

LAURIE E. McNEIL, Ph.D., is an associate professor in the Department of Physics and Astronomy, University of North Carolina at Chapel Hill.

RICHARD A. MESERVE, J.D., Ph.D., is a partner with the firm of Covington and Burling in Washington, D.C. He previously served as legal counsel to the President's Science and Technology Adviser (1977-1981).

FRANK M. RICHTER, Ph.D., is professor and chairman of the Department of Geophysical Sciences, University of Chicago.

ARTHUR H. RUBENSTEIN, M.D., is professor and chairman of the Department of Medicine at the University of Chicago. He chaired the 1989 Institute of Medicine study titled *The Responsible Conduct of Research in the Health Sciences* and is a member of the IOM council.

HOWARD K. SCHACHMAN, Ph.D., is a professor in the Department of Biochemistry and Molecular Biology at the University of California, Berkeley. He was formerly president of the Federation of American Societies for Experimental Biology and of the American Society for Biochemistry and Molecular Biology.

HOWARD E. SIMMONS, Jr., Ph.D., is vice president, Central Research and Development Department, E.I. du Pont de Nemours and Company, Inc., in Wilmington, Delaware.

ROBERT L. SPRAGUE, Ph.D., is a professor in the College of Medicine and is the director of the Institute for Research on Human Development at the University of Illinois, Urbana-Champaign.

SHEILA WIDNALL, D.Sc., is associate provost and Abby Rockefeller Mauze Professor of Aeronautics and Astronautics, Massachusetts Institute of Technology. Dr. Widnall was formerly president of the American Association for the Advancement of Science.

PATRICIA K. WOOLF, Ph.D., has conducted research and written extensively about misconduct and the sociology of science. She is a lecturer in the Department of Molecular Biology at Princeton University.

KEITH R. YAMAMOTO, Ph.D., is professor and vice chairman of the Department of Biochemistry and Biophysics at the University of California, San Francisco.

Study Staff

ROSEMARY CHALK, study director, previously directed the 1989 Institute of Medicine study on scientific conduct. She also served as program head of the AAAS Office of Scientific Freedom and Responsibility from 1976 to 1986.

BARRY D. GOLD, senior staff officer, was previously senior program associate at the AAAS. He also served as a staff officer for the AAAS-American Bar Association National Conference of Lawyers and Scientists.

DAVID H. GUSTON, research associate, is a Ph.D. candidate in the Department of Political Science at the Massachusetts Institute of Technology. He has previously served as a research assistant at the AAAS and at the Congressional Office of Technology Assessment.

B

Subpanels

SUBPANEL ON THE ENVIRONMENT OF THE STUDY

EDWARD E. DAVID, JR. (*Chairman*), EED, Inc.
ALBERT BARBER, University of California, Los Angeles
MICHAEL BERMAN, The Duberstein Group, Inc.
JOHN DEUTCH, Massachusetts Institute of Technology
BERNARD GERT, Dartmouth College

Staff

Barry Gold, Senior Staff Officer

SUBPANEL ON RESEARCH PRACTICES AND STANDARDS

IRA J. HIRSH (*Chairman*), Washington University
PHILIP H. ABELSON, American Association for the Advancement
    of Science
VICTOR R. BAKER, University of Arizona
VAL L. FITCH, Princeton University
MARYE ANNE FOX, University of Texas at Austin
PETER GALISON, Stanford University
JENNY L. McFARLAND, Massachusetts Institute of Technology
LAURIE E. McNEIL, University of North Carolina at Chapel Hill

RICHARD A. MESERVE, Covington and Burling
KEITH R. YAMAMOTO, University of California, San Francisco

Staff

David H. Guston, Research Assistant

SUBPANEL ON MISCONDUCT AND INTEGRITY IN SCIENCE

HOWARD E. SIMMONS, JR. (*Chairman*), E.I. du Pont de Nemours and Company, Inc.
FRANK M. RICHTER, University of Chicago
ARTHUR H. RUBENSTEIN, University of Chicago
HOWARD K. SCHACHMAN, University of California, Berkeley
ROBERT L. SPRAGUE, University of Illinois at Urbana-Champaign
SHEILA WIDNALL, Massachusetts Institute of Technology
PATRICIA K. WOOLF, Princeton University

Staff

Rosemary Chalk, Study Director

Index